THE MAINTENANCE PLANNERS PLAYBOOK

Chris Diprose

A Planner HQ Publication

PLANNER HQ

TOOLS AND TRAINING FOR PLANNING EXCELLENCE

First published 2025

First Edition

ISBN: 978-1-9192306-0-3

For more resources, visit: www.theplannerhq.com

Follow us on LinkedIn: https://www.linkedin.com/company/the-planner-hq/

About Planner HQ

Planner HQ is a professional development platform built to elevate the role of maintenance planners and those working at the heart of reliability, asset management, and operations.

The goal is simple: to empower maintenance planners to plan better, lead better, and deliver meaningful impact in their organisations. Through practical tools, in-depth training, and thought leadership grounded in real-world experience, Planner HQ helps individuals and teams move beyond reactive maintenance and into structured, proactive, and data-driven execution.

The project was created by people who have lived the planner role across industries, across continents, and across systems - from pharmaceutical plants and manufacturing lines to national healthcare systems and major transportation infrastructure. That real-world foundation is what makes Planner HQ different. It's not just theory - it's about what works.

Whether you are an individual planner looking to sharpen your skills, a leader seeking to upskill your team, or an organisation aiming to improve reliability outcomes, Planner HQ is designed to support your journey.

This book is one piece of a broader mission. To learn more, explore future tools, and stay connected with the evolving Planner HQ ecosystem, visit us at:

→ www.ThePlannerHQ.com

Foreword

When I first stepped into the world of maintenance planning, I had no idea how critical - and how misunderstood - the role truly was.

I quickly discovered that planners sit in a unique position: not quite engineers, not quite technicians, not quite managers - but somehow responsible for holding everything together. A good planner doesn't just schedule work; they create structure in chaos, clarity in uncertainty, and reliability in the long term. Yet despite this, the role is often undervalued, undertrained, or overlooked entirely in many organisations.

Over the years, I've had the privilege of working in some of the most demanding environments around - high-volume pharmaceutical manufacturing, large-scale logistics at Amazon, and cross-border operations supporting multiple languages, systems, and teams. I've been part of setting up Enterprise Asset Management programs at multiple international airports, rail networks, and national hospitals in Tanzania. Each of these projects came with its own pressures, politics, and complexity - but one thing remained constant: **the quality of planning directly influenced the quality of results**.

In every site, the difference between firefighting and flow was rooted in how well the planning process was respected, structured, and executed. And more importantly - how well the planner was supported.

That's why I decided to write this book.

Not as a textbook, not as a recycled list of best practices, but as a hands-on, experience-based guide to mastering one of the most high-leverage roles in modern industry. This playbook is built from the ground up with practicality in mind. If something worked in the field - it's in here. If it didn't - it's not.

Whether you're a new planner trying to find your feet, a senior planner looking to tighten up your systems with fresh ideas, or a maintenance leader building structure into your team - this book is for you. It's written with deep respect for the craft, and even deeper respect for the people doing the work.

It's also the launchpad for something bigger: **Planner HQ**.

Planner HQ is more than just a brand name - it's a growing platform built to give maintenance planners the tools, training, and community we've long been missing. While maintenance planning often happens behind the scenes, it deserves to be front and centre in the reliability conversation.

I believe planning is a craft worth mastering. I believe planners deserve recognition, development, and clear paths to growth. And I believe that when planning is done right - everything else gets better: uptime, efficiency, safety, morale, and culture.

So consider this your invitation - not just to read, but to rethink what planning can be. Dive in with curiosity, challenge what doesn't fit, share what helps your team. Above all, use this book to move one step closer to becoming the type of maintenance planner every site wishes they had.

Chris Diprose

Founder, The Planner HQ

Table of Contents

PART I – Foundations of Planning

Chapter 1

The Role of a Maintenance Planner

"You're not just planning work - you're shaping reliability"

A good maintenance planner doesn't just plan work - they create clarity, stability, and flow in a world of chaos.

Walk into any industrial facility - a distribution centre, a manufacturing plant, a hospital, or even a theme park - and you might notice hugely complex systems running behind the scenes. Pumps hum. Conveyors glide. HVAC units regulate. Robots and humans work in tandem. And none of it happens (for very long) without effective maintenance. However, good maintenance doesn't just happen - and that's where the Reliability Maintenance Planner role comes in.

Note that I've used the term "Reliability Maintenance Planner" here in the intro. This is because throughout this book I want you to be thinking with a Reliability focused mindset - that is ultimately the key output to our role and what we should be planning for. Of course, different organisations will use different job titles - commonly I've seen Reliability Planner, Maintenance Planner, Engineering Planner and Works Coordinator, to name a few popular ones. We'll generally use the straightforward term "Planner" throughout this book for simplicity.

The planner is the often-overlooked force that keeps the entire system from falling apart - literally and figuratively.

The Silent Backbone of Reliability

In simple terms, a maintenance planner should be responsible for defining, organising, and preparing maintenance strategies in a way which allows technicians and engineers to execute it safely, efficiently, and with minimal disruption to operations. But all of that is just scratching the surface.

A good maintenance planner is a:

- **Reliability centred thinker**: As touched on earlier, we need to base decisions and strategy with Reliability in mind. They steer the team away from just going through the motions of maintenance tasks, and towards a focus on equipment reliability and uptime for customers.

- **Coordinator of chaos**: They bring order to competing priorities, parts shortages, and personnel constraints. Large data sets are brought under control and presented in coherent ways.
- **Gatekeeper of quality work**: They ensure that jobs are scoped out correctly before anyone picks up a wrench. They are gatekeepers of data quality and maintenance record keeping.
- **Bridge-builder**: They connect engineering, operations, procurement, safety, and the shop floor in a role which often spans departments like no other.
- **Strategist**: They work weeks ahead of the immediate firefight, building a buffer of planned work and a stable schedule. They consider full asset lifecycles, not just the next job to be done.

What Planning ***Is Not***

One of the first hurdles for new or developing planners is clearing up misconceptions. Planning is not:

- Just printing work orders and emailing reminders
- An admin-only or “desk job” role
- About fixing today’s breakdowns

In fact, a planner should spend very little time dealing with emergencies. The moment you find yourself chasing breakdowns all day, you’re not planning - you’re firefighting, which is never going to bring about a Reliability focused team. All the time a planner is spending energy on something happening right now, you can’t be building a better and more coherent longer-term plan. There should be other team members who focus on the here and now firefighting.

Planning has to be a **proactive** role. It's about preparing tomorrow's work today, so that when technicians hit the floor, they have everything they need: clear job steps, correct parts, tools, permits, equipment availability and time allocated on the schedule.

The Value Planners Create

When done right, effective maintenance planning improves:

- **Equipment reliability** (uptime)
- **Wrench time** (actual hands-on-tool work)
- **Schedule compliance**
- **Labour efficiency**
- **Safety standards**

In my own experience, well-planned work takes 30–50% less time to complete compared to reactive work. That's not a small improvement - that's a productivity multiplier.

Think of it this way: if a technician spends 3 hours servicing a conveyor belt without a plan, but could have done it in 1.5 hours with a proper plan and appropriate preparation, that difference scales across an entire site or network. Now you're talking about millions in savings - not to mention less overtime, fewer mistakes, and happier teams.

At Amazon there is a commonly used phrase called "Customer Obsession". Think about who your customer is in these situations, and how can the work you do be obsessed with improving conditions and outcomes for your customers.

- Operations teams - how can you improve their equipment availability and minimise downtime for them? How can you enhance communications between departments, and provide visibility of upcoming maintenance?

- Technicians and Engineers - how can you reduce friction in their day-to-day job, and allow them to execute world class maintenance in the best possible way?
- Procurement Departments - how might your well-considered plans be able to reduce spare part inventory holding costs? How could streamlined job plans reduce the need for spend on third party contractors?

Always obsess over your customer as the key principle in your approach, and build things up from there.

Why is Maintenance Planning more valuable than ever?

In decades past, maintenance was often seen as a necessary evil - a cost centre, called upon only when something broke. In today's landscape, that mindset is no longer acceptable. Maintenance is now recognised as a core driver of safety, efficiency, and profitability. And at the heart of that shift is the maintenance planner.

Let's look at *why* the planner's role is becoming so vital across modern industries.

1. The Cost of Downtime is Rising

Whether you're in manufacturing, logistics, pharmaceuticals, or utilities; unplanned downtime is more expensive than ever. In many modern facilities:

- A minute of downtime on a production line can cost thousands in lost output.
- For logistics operations, downtime delays shipments, damages customer trust, and creates ripple effects in the supply chain.

- In pharmaceuticals or utilities, downtime can result in regulatory violations or health and safety risks.

Every minute saved by a good planner translates directly to money saved - or money earned, and companies have been rapidly waking up to that fact.

2. Assets Are Becoming More Complex

Today's equipment is smarter, more integrated, and often more delicate than in the past. Variable frequency drives, SCADA-controlled systems, complex robotics, and IoT-connected assets are increasingly common.

What does this mean for Planners?

- Job plans must account for electrical, mechanical, automation, and digital elements
- The technician's skill mix is more diverse and needs to be managed and utilised accordingly
- Planning can no longer be generic - it has to be precise and multidisciplinary

A planner is one of the few roles in the engineering and asset management space who is expected to understand the full picture and coordinate across disciplines.

3. Workforces Are Changing

Industries are facing a skills gap:

- Many senior technicians are retiring.
- Newer hires may lack deep hands-on experience.
- Contractors are used more frequently, and they rely heavily on clear scopes and documentation.

Planners are crucial in this environment because they:

- Transfer tribal knowledge into structured job plans.
- Enable less-experienced technicians to perform confidently.
- Standardise best practices across shifts, teams, and sites.

4. Lean Operations Leave Little Room for Error

Companies today strategically run leaner operations than ever, and this affects the engineering and maintenance departments approach:

- Fewer spares inventory on hand - fine tuned stock holding strategies required
- Less buffer time in production schedules - harder to allocate maintenance time on equipment
- Tight margins and faster delivery promises - greater consequences to unplanned downtime

This puts enormous pressure on maintenance teams to execute efficiently. Planners help this by:

- Ensuring right-first-time execution.
- Reducing rework, delays, and part shortages.
- Making maintenance predictable rather than reactive.

In many operations, it's not just about getting the job done - it's about getting it done **at the exact right moment**, without disrupting flow.

5. Digital Transformation Needs Human Structure

CMMS, IoT, Predictive Analytics, AI - the new tools of modern maintenance are evolving fast. But tools don't replace processes; they enhance it - if used well.

Planners are a key facilitator in:

- Translating digital insights into scheduled action.
- Structuring workflows inside the CMMS
- Managing data quality (bad data = bad decisions).

- Driving adoption and consistency across teams.

Put simply: without planning, all the technology in the world won't make a site reliable.

6. The Planner Role Enables Strategic Reliability

Many organisations now have formal **reliability programs**, often led by Reliability Engineers (REs) or Maintenance Managers. These programs aim to reduce reactive work, extend asset life, and control lifecycle costs, but ultimately increase equipment availability.

Planners are typically the operational arm of these reliability programs and initiatives:

- They clean up and structure asset maintenance and performance data for analysis.
- They operationalise the PM strategies developed by Reliability Engineers.
- They support Root Cause Analysis (RCA's) by deep diving and presenting detailed data, typically from the organisations CMMS.
- They simplify chaos, so reliability teams can focus on improvement.

Without planners, reliability initiatives often stall - because no one has the bandwidth to turn ideas into action and bridge the gap between the highly technical Reliability Engineers and the routine maintenance execution field teams.

The Planner's Place in the Organisation

Maintenance planners often sit in a quite unique position in an organisation. You're not a technician, but you need at least some foundational technical understanding. You're not an engineer, but you need to interpret technical drawings. You're not usually a people manager, but you need to influence teams daily to be effective. You're a systems thinker, a translator, and a conductor - making sure the right notes are played at the right time.

By way of example, at Amazon, planning is deeply embedded into the maintenance strategy. "Reliability Maintenance Engineering Planners" are expected to:

- Own the weekly planning and scheduling process
- Drive CMMS data quality and PM optimisation
- Partner with procurement to ensure spare part availability and budget adherence
- Serve as a primary point of contact between operations and maintenance
- Analyse work order backlogs and drive continuous improvement

Even in smaller organisations, the core responsibilities are the same - the scale and tools just change.

A Role That Demands Range

Here's what the best planners I've worked with have in common:

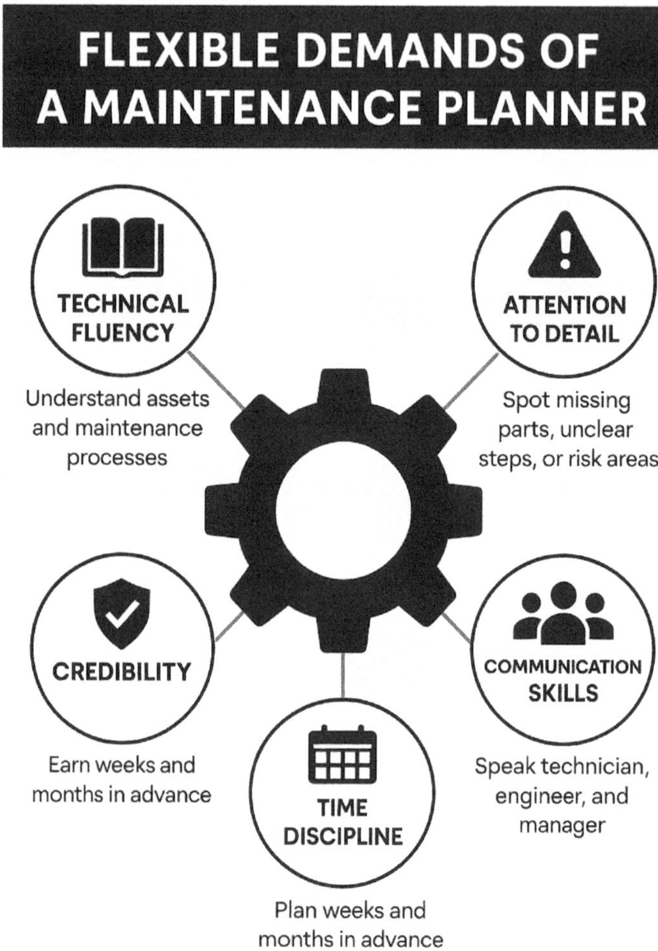

- **Technical fluency**: They understand assets and maintenance processes. They may not be technically qualified or apprentice served, but they have a good all-round understanding of the equipment under their scope and can sufficiently converse on that level.

- **Attention to detail**: They spot missing parts, unclear steps, or risk areas before the work begins. They insist on the highest standards of data quality and record keeping, recognising the downstream value that comes with it, and the downside risks of poor data quality standards.
- **Communication skills**: They speak technician, engineer, and manager; and can seamlessly be the bridge between engineering and operational leadership. They are able to sell the value in their strategy in order to gain the trust of the teams around them.
- **Time discipline**: They plan weeks and months ahead, not hours. Long term is where reliability is truly achieved, not short term.
- **Credibility**: They've earned the trust of the trades - because they've walked the walk or worked to understand it. Anything they don't understand they are honest about, and don't pretend to know something they don't.

You don't have to be the most experienced technician to be a great planner. But you do have to respect the craft, listen to your technicians, and constantly seek better ways to support them.

The Bottom Line

The maintenance planner role is no longer "nice to have." In most modern, asset-intensive organisations, it is a force multiplier. It allows:

- Fewer technicians to achieve more.
- More work to be done in less time.
- Higher uptime, safer conditions, and lower cost per repair.

And as more companies adopt leaner models, push for digital integration, and fight for operational excellence, **the planner's role will only grow in value**.

If you are in a maintenance planner role already, you are in the right place. If you're aspiring to get into it - the timing couldn't be better!

This book is written for those who want to become indispensable in their Planner role. Whether you're just stepping into planning or you've been doing it for years and want to level up, I'm confident you'll find tools, strategies, and hard-won lessons throughout these pages.

Being a great maintenance planner isn't about working harder. It's about thinking clearer, preparing better, and executing smarter. That's what separates the average from the world-class; and if you're reading this, I'm guessing you're aiming for the latter.

Key Takeaways

- Maintenance planning is not an admin task-it's a strategic function that directly impacts reliability, safety, and operational performance.
- A great planner is a reliability thinker, process coordinator, data gatekeeper, relationship builder, and long-term strategist.
- The planner's primary customer isn't just the technician-it's also operations, procurement, and the wider business.
- Planning must be proactive. If you're constantly reacting to breakdowns, you're not truly planning-you're firefighting.
- Effective planning enables higher wrench time, better job execution, safer work, and significant cost savings.
- Today's planners must navigate complex equipment, digital systems, and evolving workforces-all while driving structure and clarity across departments.
- The modern planner bridges gaps between teams, systems, and strategies-translating reliability goals into real, executable work.
- Your value as a planner grows with your ability to prepare, communicate, and lead without formal authority.
- As businesses go leaner and more digitally integrated, planning isn't optional-it's essential.
- This book will help you elevate your planning practice, become a force multiplier in your team, and lead the way to greater reliability.

Chapter 2

The Skills of a Great Maintenance Planner

"Influence isn't about authority - it's about preparation, consistency, and trust."

A great maintenance planner is more than an administrator-they are a translator of technical language, a strategist, a data analyst, and above all, a leader through influence. The best planners don't just move work through a system; they shape how work gets done, who does it, and how well it's executed.

In this chapter, we'll explore the core skills that elevate a planner from good to great, focusing on technical know-how, systems thinking, high-performance mindsets, and influence.

1. Technical Know-How

Maintenance planning starts with a foundational level of technical understanding. A planner needs enough equipment knowledge to speak confidently with technicians, recognise bad work requests, and build job plans that make sense. You don't need to be the best tradesperson in the room-but you do need to know what a good job looks like, and you need to be able to engage with technically minded people.

Key Technical Foundations:

- **Understanding Equipment**: Know how critical assets function, their key components, and how they interact with surrounding systems. This doesn't mean deep technical design knowledge, but enough to understand common failure modes and what "normal" should look like. If you are new to an organisation or industry - take time to learn the operational process from the start, to fit your plans seamlessly into that world.
- **Maintenance Methods**: Be well-versed in preventive maintenance (PM), condition-based monitoring (CBM), non-intrusive System Check (SC), and corrective maintenance (CM). This helps you match the right strategy to the task, rather than trying to use 'one size fits all' approaches.

- **Tools and Parts**: Understand the tools, parts, and consumables required for standard jobs. A good planner anticipates technician needs, ensuring everything is ready for efficient execution.

Technology Proficiency

The modern maintenance planner is very much a digital worker. While we will dive deeper into the systems side of planning in later chapters, a strong baseline in software tools is non-negotiable:

- **CMMS (Computerised Maintenance Management Systems)**: This is your life-blood. You'll use this daily - for managing work orders, planning jobs, tracking history, and reporting KPIs.
- **Excel / PowerBI (or other data analysis software)**: Critical for analysing trends, manipulating and presenting data, building templates, and even bridging gaps where systems fall short.
- **Project Planning Tools**: Software like MS Project, Primavera, or even simple Gantt tools are useful for shutdown planning or long-term maintenance projects.
- **Collaboration Software**: Planners often serve as the hub between maintenance, operations, and procurement. Tools like Microsoft Teams, Slack, Quip, Asana, or many other shared project boards help manage communication and transparency across departments.

Think of technical know-how not just as "hands-on" ability, but as an understanding of how to make technicians' jobs easier - through better preparation, better systems, and better support.

2. Systems Thinking

A great planner doesn't just see a job-they see a system. Systems thinking means understanding how every piece fits together: the machine, the supply chain, the work process, and the people. This skill allows planners to go beyond reactive fire-fighting and begin designing processes that make the maintenance strategy more proactive, consistent, and scalable.

Core Elements of Systems Thinking:

- **Holistic View of Maintenance**: Recognise that equipment failures aren't just technical - they're often process failures. Missing spare parts, poor communication, unclear job scopes, below par maintenance execution; these are symptoms of broken systems.
- **Feedback Loops**: Great planners build feedback loops into their work - reviewing completed jobs, using technician feedback, and updating plans continuously to improve future performance.
- **Workflow Design**: A planner who understands workflows - from work request to completion - can identify and eliminate friction, automate repetitive steps, and improve visibility.

Systems World-Class Planners Use:

- **Standard Job Plans and Libraries**: Reusable plans reduce planning time, improve consistency, and standardise expectations (such as those developed by The Planner HQ)
- **Planning Checklists and Templates**: These help maintain quality across tasks and reduce the mental load of starting from scratch.
- **Backlog Management Structures**: Backlogs organized by asset, priority, and craft help teams pull the right work at the right time.
- **Work Week Structures**: Systems like "Frozen Schedules" or "Rolling 2-Week Planning" create predictability, reduce disruptions, and increase efficiency.

A systems-minded planner doesn't just make things run - they make them run better over time. By proactively identifying recurring pain points and designing workflows that reduce waste and ambiguity, planners elevate themselves from task managers to system optimisers. The result is fewer breakdowns, less stress, and a smoother operation across the board.

3. Mindsets of a High-Performing Planner

Mindset is what transforms knowledge into consistent execution. While skills can be learned, mindset must be cultivated. The most effective planners embody three powerful mental frameworks: proactivity, data-driven thinking, and disciplined consistency.

Proactive by Nature

Great planners constantly look ahead. They anticipate needs, identify potential blockers, and build buffers into the schedule.

- Ask yourself: *What can I plan now to prevent a failure later? What system can I build to reduce process errors in future?*
- Focus on building a healthy backlog and maintaining tight scheduling discipline.
- Develop the habit of preparing - whether it's parts kitting, updating BOMs, or refining job steps.

Data-Driven Thinking

Planners sit on a goldmine of data - work order histories, failure reports, completion rates, and backlog metrics. The best use this information to drive better decision making:

- Prioritise work based on frequency, downtime impact, or cost.
- Identify recurring failures and plan systemic solutions.

- Use KPIs like schedule compliance or mean time between failure (MTBF) to measure planning effectiveness.

Example: If 40% of your PMs result in corrective work, dig deeper. Are the PMs poorly written? Have all steps been carried out or shortcuts taken? Is the equipment nearing end-of-life? Data shows you where to ask better questions.

Discipline and Consistency

Planners are the heartbeat of the maintenance schedule, and that heartbeat must be steady.

- Review work requests daily.
- Update your backlog weekly.
- Communicate status reports regularly for complete transparency.
- Hold to a schedule, even when it's uncomfortable.
- Share long term plans and maintenance strategy openly.

This discipline and consistency builds trust - and when trust grows, so does your influence.

4. Influence: Leading Without Authority

The final and often most overlooked skill of a world-class planner is influence. You don't manage technicians, control production, or run procurement; but your role depends on all of them. Your ultimate success largely depends on your ability to influence across the organisation, without formal authority or direct line reporting structure.

Communication:

Influence starts with clear, consistent communication.

- **Write Clear Job Plans**: Clarity reduces ambiguity, mistakes, and frustration.
- **Hold Regular Planning Meetings**: Weekly scheduling meetings with operations build alignment and reduce last-minute surprises.
- **Share Your Strategy:** Don't hide your long-term plans in a folder. Make them visible, engage your team in fine tuning them, and raise their profile.
- **Close the Loop**: Follow up on completed work. Ask yourself and your team: *Did the job go to plan? What can we improve? What did we try that didn't work well?*

Building Relationships:

Trust is the currency of influence. Invest heavily in relationships across the plant.

- **With Technicians**: Take time to walk jobs, ask for feedback, and acknowledge their experience. A respected planner gets better information and better results.
- **With Operations**: Learn their pain points and align your priorities with theirs as much as possible. Offer solutions, not problems. *Remember: Customer Obsession*
- **With Stores/Procurement**: Understand spare part lead times and stock levels. Help forecast demand to avoid parts delays. Collaborate with procurement on sourcing of parts and third-party vendor support.

Influence doesn't stop at the shop floor. World-class planners also learn to present value to senior leaders. Use your data to show impact: reductions in reactive work, improved schedule compliance, fewer breakdowns. Tie your planning improvements to uptime, cost savings, and team morale. Leaders support what they can see - so help them see the power of good planning. We will explore the importance of building influence in more detail through chapter 16, but it's important to keep the concept in mind as we progress through this journey.

Practical Advice to Build Influence:

- Be present on the floor. Visibility builds familiarity and trust.
- Deliver on commitments. If you say parts will be ready - make sure they are.
- Share wins. Celebrate when schedules run clean, jobs close out well, or technicians provide great feedback.
- Ask questions, not just for answers - but to show people you value their input.

You don't need authority to lead - you need consistency, communication, and a trusted reputation for getting things done.

Owning the Process

One of the most powerful ways a planner builds influence is by becoming the person others turn to for clarity, structure, and consistency.

When you deeply understand how your teams maintenance work flows - from work intake to closure - you can spot where it breaks down and take ownership of making it better. People naturally follow those who improve their day-to-day work.

- Own the weekly planning and scheduling cycle. Bring the discipline, and people will align with you.
- Use your CMMS better than anyone else. Show how it can answer questions and prevent chaos. Position yourself the team Subject Matter Expert (SME) for your CMMS and you'll have much more influence on how it is used.

- Track key metrics - work order backlog health, schedule compliance - and explain what they mean to the team. Metrics need to be understood and visible for them to add value.
- Document processes that others take for granted, and be the one who simplifies complexity.

The more structure you bring, the more people look to you for answers. And when that happens, your influence becomes embedded-not because of your title, but because of your value.

We can see then, that the modern maintenance planner wears many hats - and none of them are optional. To succeed in the role, you must combine technical awareness with process insight, a proactive mindset with data fluency, and strong communication with the subtle art of influence.

- Understand the equipment well enough to speak confidently with technicians and create reliable job plans.
- Think in systems - seeing beyond the job at hand to the workflows, structures, and strategies that support it.
- Operate with a mindset of discipline, proactivity, and data-driven thinking, ensuring their actions are consistent and defensible.
- Build influence not just through relationships, but by owning the planning process and becoming the calm centre of the maintenance machine.

You don't need to be the loudest voice in the room. When you master these skills, your work speaks for itself and others will begin to follow your lead. That's the path to becoming a trusted, indispensable maintenance planner in any organisation.

Key Takeaways

- **Technical know-how matters**-you don't need to be the best technician, but you must understand how equipment works, what good maintenance looks like, and how to build job plans that actually help.
- **Know your tools**-from CMMS to Excel to project planning software, your digital skills shape how efficiently you manage, communicate, and track work.
- **Think in systems**-see beyond individual work orders and understand the end-to-end maintenance process, from backlog health to spare part flow to workflow friction points.
- Use **templates, libraries, and planning structures** to create consistency and reduce rework across the team.
- **Adopt a proactive mindset**-look ahead, build buffers, refine schedules, and reduce firefighting through preparation and process control.
- **Let data guide you**-track what's working, what isn't, and where failures repeat. Use KPIs to measure planning performance and improve strategy over time.
- **Be disciplined and consistent**-reliability depends on a steady cadence of review, planning, and follow-through.
- **Influence is your superpower**-lead through clarity, communication, and credibility, even if you don't have formal authority.
- Build trust across departments-technicians, operations, and procurement-and they'll start pulling in your direction.
- **Own the planning process**-when people trust your systems, data, and foresight, they'll turn to you for answers. That's when real influence takes root.

PART II – Core Practices and Tools

Chapter 3

The Maintenance Planning Process

"A planned job starts long before the wrench turns - and doesn't end when the work is done."

Maintenance work doesn't start when a technician grabs their toolbag - and it doesn't end when the job is marked complete and the work order is closed off. Behind every well-executed maintenance activity is a structured process that moves work from identification to execution. As a planner, your role is to guide and optimise that process - to remove friction, maximise clarity, and ensure each task contributes to a more reliable facility.

This chapter breaks down the full maintenance planning process, showing where and how a planner adds value. It also defines the crucial difference between planning and scheduling, and why this distinction is key to reliability-centred maintenance.

The Lifecycle of Maintenance Work

Every maintenance job follows a basic lifecycle. When the process is healthy and well-managed, work flows smoothly, field teams are productive, and equipment reliability should improve as an end result. When it's broken or chaotic, the opposite happens - technicians chase breakdowns, parts arrive late, and the same failures repeat - equipment becomes unreliable.

Here's a high-level view of the typical maintenance work lifecycle:

- Work Identified
- Work Request Created
- Work Order Approved and Prioritised
- Work Order Planned
- Work Scheduled
- Work Executed
- Work Closed and Documented

Let's walk through each stage-and how the planner contributes value throughout.

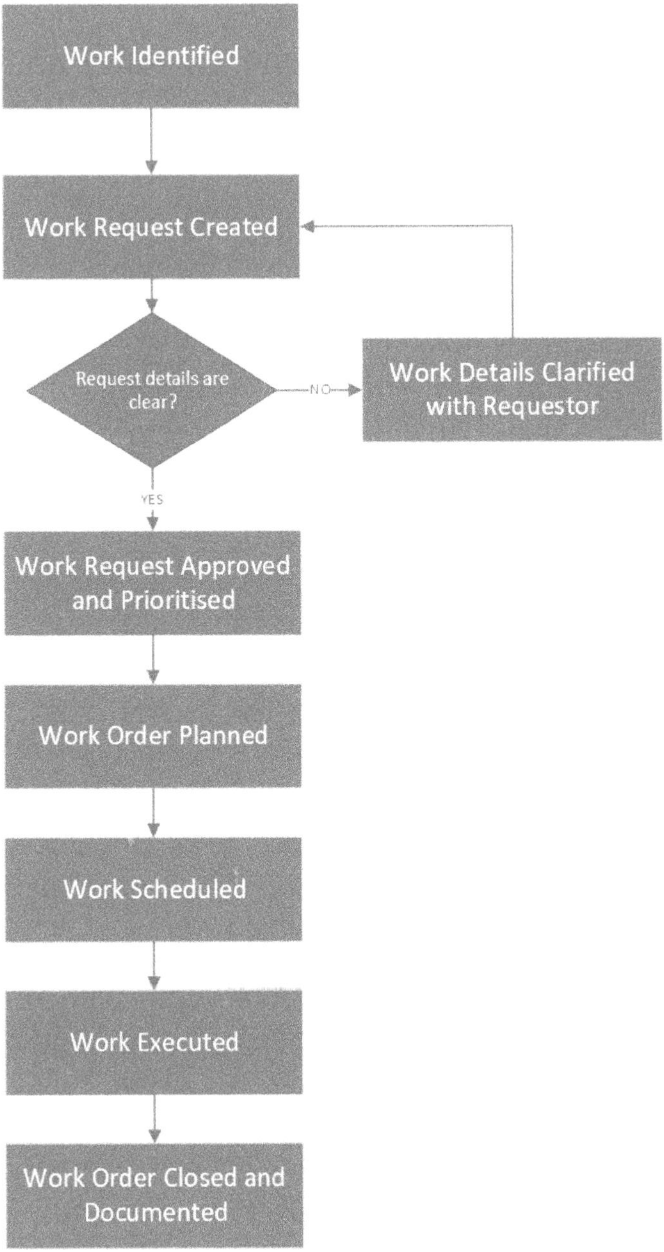

Figure 1 - The Lifecycle of Maintenance Work Requests

Work Identified

This is the trigger. A (potential) issue is noticed - through inspection, condition monitoring, or someone on the floor reporting an issue.

Planner's Role: While planners may not be directly identifying work, they have a key role in enabling good identification practices. This includes supporting robust inspection programs, ensuring operators and technicians know how to raise and log work requests, usually into the CMMS.

They champion and help develop robust systems like System Health Checks, Condition Monitoring or Predictive Maintenance routines to best facilitate opportunities to identify issues before they develop into a full equipment breakdown.

Planners also help close the loop by tracking failure trends and initiating new proactive tasks based on repeated issues. In a reliability-centred culture, planners are not just administrators - they are data stewards and analysts who help steer attention toward emerging risks.

Work Request Created

A work request is submitted into the CMMS via one of the routes described in the last step. The quality of this request matters greatly - a vague or incomplete submission leads to wasted time and guesswork.

Planner's Role: Triage and validate every incoming request. A world-class planner reviews each one to ensure it includes key details: what's wrong, where it is, how it was noticed, and any immediate impact.

If a request is unclear, the planner follows up directly with the requester. This is also where duplicate requests or non-maintenance issues are filtered out. A strong gatekeeping process here protects the planning team's time and sets the foundation for productive work. It also ensures only high quality and crystal clear work requests can reach the field engineering team, something which will go a long way in your endeavour to earn trust to gain influence - something explored in more detail later.

Work Order Approved and Prioritised

Once validated, requests can be processed into work orders and assigned a priority. This could be done directly by the planners, or in some organisations this task moves over to maintenance or operations leadership, but planners should always play an essential advisory role in the prioritisation.

Planner's Role: Provide data and context to help leaders make smart prioritisation decisions. For example:

- Has this issue occurred before?
- What's the impact on safety, quality, or uptime?
- Are parts readily available?
- What other works are planned/in progress?

Planners may also identify if the work ties into larger projects or existing PMs, helping avoid duplication. A planner with a strong systems view ensures the backlog remains aligned with overall business priorities - not just whoever shouts the loudest.

Work Order Planned

This is where the planner shines. Turning an approved work order into a well-prepared, executable task is the core of the planning role.

Planner's Role:

- **Define the Scope**: Clarify what work is in and out of scope. This avoids ambiguity later.
- **Break Down Tasks**: Identify the logical steps to complete the job.
- **Identify Resources**: Determine trade types, headcount, skill levels.
- **Estimate Time**: Include setup, teardown, testing, and documentation.
- **List Tools and Parts**: Specify everything needed, ideally from standard service kits or BOMs.
- **Stage Materials**: Work with stores to ensure materials are available. If possible, pre-kit and stage them.
- **Attach Documents**: Include drawings, procedures, permits, and safety notes as applicable.

The overall goal is to create a job plan that allows a technician to complete the work without needing to improvise, search for parts, or stop and ask for clarification.

Software Tools: Planners must also be fluent in the digital tools of the trade:

- **CMMS**: The backbone of planning-used to track work orders, parts, labour, and asset history.
- **Excel** for templates, parts lists, and standard work libraries.
- **Project Planning Tools** (e.g. MS Project, Smartsheet) for more detailed shutdown or campaign planning.
- **Collaboration Software** (e.g. SharePoint, Teams, Quip) for document sharing and version control.

To excel as a planner you don't need to be an IT expert - but you do need to be confident navigating these systems to keep the planning process clean, visible, and consistent.

Work Scheduled

Planning defines what needs to be done. Scheduling determines when it will happen. And in many organisations, it's the planner who leads the charge in proposing that "when."

Planner's Role: A great planner doesn't just drop jobs into a backlog - they build a forward-looking, achievable schedule. This means assessing the time each job will take, matching it against available technician hours, and proactively working with operations to secure the right equipment access windows (planned downtime)

While the final schedule might be approved or adjusted by supervisors or operations, the planner sets the baseline. You're the one who:

- Proposes realistic timing based on estimated duration, resource capacity, and backlog health
- Flags work that must be grouped for efficiency (e.g. shared equipment access or technician specialty)
- Requests access windows and downtime from operations
- Prepares the weekly plan that becomes the basis for scheduling meetings
- Keeps scheduling grounded in real capacity, not just urgency

A planner-led schedule reduces noise. It gives the team a clear direction of work, allows for proper resource/material preparation, and maximises wrench time. You're not just enabling execution-you're shaping the tempo of the entire maintenance operation.

Work Executed

This is where the technician brings the plan to life. While the planner isn't the one holding the tools, their preparation is what makes the execution possible - or painful.

Planner's Role:

- Ensure all materials, spare parts, tools, and kits are available and in the right location
- Attach job plans, permits, and supporting documentation in the CMMS or at point of use
- Remain available to clarify scope or adjust job details in real-time if unexpected issues arise

The technician may turn the wrench, but the quality of their work experience often comes down to how well you've prepared. Good planning removes guesswork. It ensures the right work happens at the right time with minimal disruption.

Poorly planned jobs cause delays, confusion, and lost trust. Great ones build flow, confidence, and reliability. Your fingerprints are on every successful execution-even if you're not there to see it.

Work Closed and Documented

Closing a work order isn't just ticking a box-it's where the real value gets captured. What actually happened? What did it cost? What can we learn? This is your chance to turn execution into insight.

Planner's Role:

- Ensure technicians accurately complete close-out details:
- Completion notes (what was done and why)
- Failure codes and root cause indicators

- Actual labour time logged
- Parts and materials consumed
- Any follow-up actions or further recommendations
- Review submitted work orders for quality and completeness
- Use this data to refine job plans, task durations, and maintenance strategies
- Escalate common issues (missing parts, repeat failures, poor access) for root cause analysis

Closure is also the last chance to correct mistakes, standardise data entry, and keep your CMMS clean. Over time, good close-out discipline builds a high-quality dataset that supports smarter decision-making.

Whether it's reporting on maintenance cost per asset, identifying high-failure components, or measuring wrench time-you can't manage what isn't captured. Your job as planner is to make sure the story gets told… and told correctly.

Planning vs. Scheduling: Why It Matters

This distinction can't be overstated.

- **Planning** defines the what and how.
- **Scheduling** defines the when and who.

When these blur, reliability suffers. If you schedule unplanned work, you're asking technicians to improvise. If you plan but don't schedule, good jobs gather dust.

World-class teams sequence them properly: plan first, then schedule. This supports proactive work, increases efficiency, and builds trust.

As a planner, you are the guardian of this flow. Push back on attempts to shortcut the process - every rushed job adds risk.

The Reliability-Centred Perspective

Planning isn't just about getting work done-it's about doing the right work, in the right way, at the right time. And the right goal? Reliability.

A reliability centred planner looks beyond just throughput and backlog closure. They focus on building a system that protects asset uptime, reduces firefighting, and supports the people doing the work.

What that means in practice:

- Prioritising proactive work like preventive maintenance (PM), inspections, and condition-based tasks-because every breakdown you prevent saves time, cost, and morale
- Studying failure history data to identify patterns-then redesigning PMs, parts strategies, or job steps to eliminate repeat issues
- Building job plans that solve root causes-not just treat the symptoms
- Refining plans over time based on close-out data, technician feedback, and equipment performance trends

Reliability-centred planning asks:

- Is this work adding value?
- What failure are we trying to prevent?
- What's the most efficient way to keep this asset available long-term?

This mindset also means respecting technician time - not wasting it with vague instructions, unnecessary disassembly, or duplicate tasks. When planners embrace reliability thinking, the whole system improves: better uptime, better data, better engagement.

The maintenance planning process is more than just pushing work through a CMMS. It's a deliberate, value-adding workflow that transforms raw requests into meaningful reliability outcomes.

When each step is respected-and when the planner's role is clearly defined-the result is a more efficient, more predictable, and ultimately more reliable maintenance operation.

In the next chapter, we'll go deeper into the craft of utilising CMMS software to drive your planning processes in the best possible ways.

Key Takeaways

- Maintenance is a **full lifecycle process**, from identifying a job to closing it with accurate data-and the planner plays a vital role in every step.
- **Work Identification**: Planners support condition monitoring, inspection systems, and data trend analysis to enable early detection of issues and build a proactive culture.
- **Work Request Validation**: A planner triages every incoming request, ensuring it's clear, complete, and appropriate before it becomes a work order.
- **Prioritisation**: Planners help assess risk, urgency, and history to ensure that the most important work gets done first-and distractions don't derail the team.
- **Work Planning**: This is your core function. Define scope, estimate time, gather tools and parts, attach drawings and permits-build a plan that technicians can trust.
- **Scheduling**: Great planners lead the conversation around *when* jobs happen. They balance resources, access windows, and backlog health to create achievable, structured weekly plans.
- **Work Closure**: This is where data quality is won or lost. Ensure technicians log what happened, how long it took, and what parts were used-because smart decisions depend on this data.
- **Planning vs Scheduling**: Know the difference and defend it. Planning defines *what* gets done and *how*, scheduling defines *when* and *by whom*. Mixing them creates chaos.
- **Reliability-Centred Thinking**: Always link planning work to the bigger picture-preventing repeat failures, improving uptime, and supporting long-term performance

Chapter 4

Introduction to CMMS

"A CMMS is only as smart as the people and processes behind it"

Maintenance planning in today's industrial environment cannot succeed without a deep understanding and effective use of a Computerised Maintenance Management System (CMMS). This chapter takes a detailed look at what a CMMS is, what features make a system valuable, how to use it effectively, and the crucial role of data quality and user training.

What Is a CMMS?

A Computerised Maintenance Management System, or CMMS, is a software platform designed to facilitate the management and optimisation of maintenance activities within an organisation. Its primary purpose is to centralise all maintenance related data - such as work orders, asset information, inventory, labour, and scheduling - in one accessible place. By digitizing and automating these processes, a CMMS contributes to improved reliability, reduced downtime, and supports data backed decision-making.

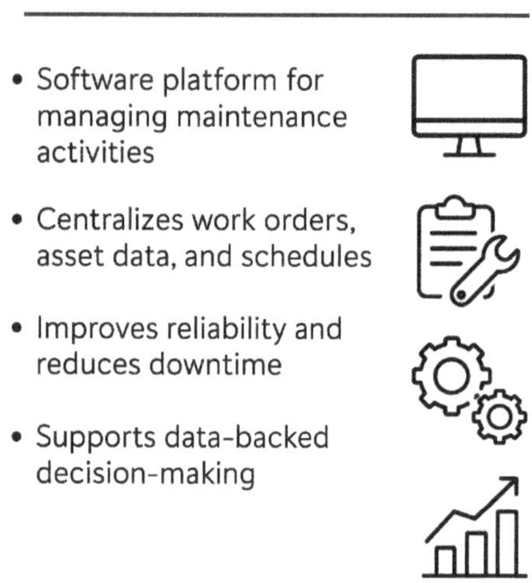

Unlike traditional paper-based or spreadsheet methods, a CMMS offers a structured, auditable, and scalable way to plan, execute, and analyse maintenance work. For maintenance planners especially, it is the essential tool to organise daily operations, ensure preventive maintenance compliance and maximise asset life.

What Makes a Good CMMS?

These days there are a whole range of CMMS platform options, developed by all sorts of companies, and the finished products do vary in effectiveness, ongoing support, and crucially; cost. Selecting a CMMS is a critical decision for any maintenance organisation. While many solutions exist, a good CMMS consistently exhibits the following qualities:

- **User-Friendly Interface**
  A CMMS should be intuitive to use for all roles - planners, technicians, managers, and procurement staff. If the interface is complicated or non-intuitive, users will resist adopting it, leading to poor data quality and lost benefits. The system should allow quick navigation, easy work order entry, and clear dashboards showing priorities and status.
- **Asset Management**
  Effective asset management means the CMMS can track detailed information about equipment, including serial numbers, location, criticality, maintenance history, and associated documentation (e.g., manuals, warranties). A good system supports hierarchical asset structures to model complex equipment assemblies, enabling planners to understand dependencies and failure impacts.
- **Work Order Management**
  The system should allow planners to create work orders with clear scopes, assign labour and materials, prioritise work (based on asset criticality or compliance requirements), and track progress from creation through completion. Automated notifications for overdue or upcoming work help maintain discipline. Processing for the field technicians should also be straightforward and easy to navigate.
- **Inventory and Parts Tracking**
  A CMMS with integrated inventory management tracks spare parts availability, usage, reorder points, and lead times. This

capability prevents downtime caused by missing critical parts and enables just-in-time inventory management to reduce carrying costs (depending on the team's inventory strategy of course).

- **Reporting and Analytics**
  Meaningful data is useless without the ability to analyse it. A good CMMS offers customisable reports and KPI dashboards - such as equipment uptime, maintenance backlog, mean time to repair (MTTR), and preventive maintenance compliance - that help planners and managers make data-driven decisions. The option to export datasets out into Excel is also a must-have for further interrogation of data.
- **Integration Capabilities**
  Modern CMMS platforms can connect with other enterprise systems like ERP (Enterprise Resource Planning), SCADA (Supervisory Control and Data Acquisition), or IoT sensors, allowing automated data capture, condition monitoring, and streamlined procurement.
- **Mobile Access**
  Field technicians can benefit from mobile apps or web portals that let them receive, update, and close work orders on the go, improving communication and timeliness.
- **Customisation and Scalability**
  Different organisations have unique requirements. A good CMMS allows tailoring workflows, fields, and reports to meet specific operational needs and scales from small plants to global enterprises.
- **Regulatory Compliance Support**
  For industries with strict safety or environmental standards, a CMMS should track inspections, certifications, and ensure documentation is complete for audits.

Examples of Popular CMMS Software

Understanding the landscape of CMMS solutions helps planners appreciate the diversity and capabilities available, and here are some of the popular options available at the time of writing:

- **IBM Maximo**
  Maximo is an enterprise-grade CMMS with a reputation for depth and flexibility. It supports complex asset hierarchies, contract management, and advanced analytics. It tend to be favoured in heavy industries like manufacturing, oil & gas, and utilities, as well as large scale operations like airports and rail networks. Maximo can integrate deeply with other IBM and third-party software, but it requires significant configuration and investment.
- **InforEAM**
  Infor's Enterprise Asset Management system balances enterprise capabilities with ease of use. It emphasises regulatory compliance, risk management, and mobile access. It supports cloud and on-premise deployment, making it adaptable to different IT strategies. A well supported industry solution, used by the likes of Amazon.
- **Fiix**
  Fiix is a cloud-native CMMS aimed at mid-sized companies wanting quick implementation and collaboration tools. Its interface is designed to be intuitive, with built-in analytics and maintenance workflows that can be customized without coding.
- **Maintenance Connection**
  A modular system allowing incremental adoption of features. It's known for strong reporting tools and an easy learning curve, making it suitable for organisations transitioning from manual systems.
- **UpKeep**
  UpKeep's mobile-first approach makes it popular for facilities

management and maintenance teams with high field technician activity. The app allows technicians to easily report status and request parts, supporting quick updates from the field.

Each platform has trade-offs in complexity, cost, and functionality. The choice should be driven by organisational size, industry requirements, and user capabilities.

How CMMS Brings Value

Transitioning from spreadsheets or paper to a CMMS yields numerous tangible benefits. While the Planners are rarely a big part in the decision making for an organisation to transfer to a CMMS, or even to upgrade, any expiring planner should be very clear on how a good and well managed CMMS can bring value when utilised correctly:

- **Centralised Data Repository**
  Instead of fragmented records in various places, a CMMS consolidates data for easier access and consistency. This reduces errors and duplication.
- **Enhanced Visibility and Control**
  Real-time dashboards give planners and managers a clear picture of workload, asset health, and maintenance effectiveness. This transparency supports proactive decision-making.
- **Improved Communication**
  Automated notifications and mobile work orders reduce delays and misunderstandings between planners, technicians, and supervisors.
- **Reduced Equipment Downtime**
  By systematically managing preventive maintenance and enabling condition-based triggers, CMMS use drives higher uptime and fewer emergency repairs.

- **Optimized Inventory Levels**
  Avoid the costs of excess stock while ensuring critical spares are available. Accurate inventory tracking prevents costly delays.
- **Data-Driven Continuous Improvement**
  Historical data collected by the CMMS enables analysis of failure patterns and maintenance effectiveness, fostering informed reliability improvement initiatives.
- **Audit and Compliance Readiness**
  A well-maintained CMMS supports traceability and documentation needed for regulatory audits or certifications.

Leveraging CMMS Tools for Efficiency

To extract maximum benefit from a CMMS, planners should actively leverage its key features:

- **Standardised Equipment Structure Hierarchy**
  Defining an organisation-wide standard hierarchy and naming conventions for your equipment data allows for accurate and comparable asset data to be utilised downstream.
- **Standard Job Plans and Templates**
  Building and maintaining a library of standard job plans ensures consistency, speeds up work order creation, and helps new technicians understand expectations. Templates can include labour requirements, materials, safety steps, and estimated times.
- **Scheduling Boards and Calendars**
  Visual scheduling tools allow planners to assign work efficiently, avoid overloading technicians, and balance urgent and routine tasks. Effective scheduling reduces downtime and improves workforce productivity.
- **Automated Alerts and Notifications**

Setting up alerts for overdue work, upcoming inspections, or low inventory helps planners stay ahead of issues and avoid last-minute firefighting.

- **KPI Monitoring and Analytics**
  Regularly reviewing reports and key performance indicators helps identify bottlenecks, uncover recurring failures, and optimize maintenance strategies.
- **Mobile Workflow Enablement**
  Encouraging technicians to update work orders in real time using mobile devices reduces paperwork and errors. This immediacy improves data quality and responsiveness.
- **Integration with Condition Monitoring and IoT**
  Linking CMMS with condition monitoring systems automates work request generation based on real-time asset health data, enabling predictive maintenance strategies that reduce unplanned downtime.

CMMS User Training and Data Quality

A CMMS's power depends heavily on the quality of data entered and the users who interact with it. Without proper training and discipline, the system can become a digital landfill of inaccurate or incomplete information, which can sometimes be even worse than having no information at all.

Training End Users (Technicians)

Technicians are the primary data inputters into the CMMS; via Work Order processing, change requests, purchase orders etc. Training should be practical and focused on daily usage:

- **Hands-On Sessions**
  Show technicians how to receive and process work orders, document progress, and close jobs using mobile or desktop interfaces. This is also a Planners opportunity to truly

engage end users with the CMMS Practice scenarios build confidence

- **Clear and Simple Job Aids**
  Provide quick reference guides or cheat sheets covering common tasks, making it easy for technicians to refresh their knowledge.
- **Ongoing Support and Feedback**
  Establish channels for technicians to ask questions or report issues. Supervisors should regularly review their input and provide constructive feedback.
- **Highlight Benefits**
  Communicate how accurate and timely data entry helps reduce repeat failures, improves safety, and makes their jobs easier by ensuring the right resources are available.

Reinforcing Data Quality on Completed Work Orders

Planners must stress to technicians the importance of recording accurate and complete information when closing work orders:

- **What Was Done**
  Detailed descriptions help planners understand the work scope and any deviations from the plan.
- **Why the Work Was Needed**
  Was it preventive maintenance, corrective repair, or a modification? Clarifying the reason informs future planning.
- **When the Work Was Done**
  Accurate timestamps support maintenance history tracking and compliance.
- **Duration and Labour**
  Actual time spent helps refine future job plans and labour estimates.
- **Materials Used**
  Correct part numbers and quantities ensure inventory accuracy and cost tracking.

- **Safety or Quality Observations**
  Any hazards encountered or quality issues observed should be documented to support continuous improvement.

Data Quality Auditing, Feedback, and Control

Maintaining high-quality data in the CMMS requires continuous effort and control:

- **Regular Audits**
  Planners should routinely review a sample of completed work orders to check for completeness and accuracy. Look for missing fields, vague descriptions, or inconsistencies.
- **Constructive Feedback**
  Feedback to technicians should be specific and respectful, explaining why accurate data matters and how it benefits the team.
- **Training Updates**
  If audits reveal recurring issues, update training materials and processes accordingly.
- **System Controls and Validation**
  Use the CMMS's capabilities to enforce data quality, such as mandatory fields, dropdown menus for standard responses, and automated validation rules.
- **Performance Metrics**
  Track data quality indicators as part of team or individual performance evaluations. Recognize and reward good practices.

The CMMS is the digital heart of a modern maintenance operation. It empowers planners to coordinate complex work, track asset health, and make informed decisions that increase reliability and reduce downtime.

However, the system's value depends on people-their knowledge, discipline, and commitment to accurate data. As a maintenance planner, mastering the CMMS and promoting its proper use is central to your role and to driving operational excellence.

Key Takeaways

- A CMMS is the backbone of modern maintenance planning, centralising work orders, asset data, inventory, and scheduling to enable structured, data-driven execution.
- Effective CMMS platforms are user-friendly, scalable, and customisable, with features like asset hierarchies, work order tracking, parts management, reporting, and integration with other business systems.
- For planners, the CMMS is more than a record system-it's a planning engine. Leveraging tools like job plan templates, scheduling boards, alerts, and mobile access enhances both efficiency and quality.
- Data quality is critical. A poorly maintained CMMS can create confusion, while clean, consistent data unlocks real insights, better decisions, and continuous improvement.
- Training and discipline are non-negotiable. Technicians must be trained to input accurate information, and planners should audit work orders regularly to reinforce standards.
- The CMMS supports regulatory compliance, spare parts optimisation, and performance tracking, all of which contribute to improved reliability and reduced downtime.
- Ultimately, the value of the CMMS is unlocked through human behaviour. Your ability as a planner to promote adoption, train users, and maintain data integrity directly influences how powerful the system becomes.

Chapter 5

Equipment Structures and Asset Metadata

"Good structure today means smarter decisions tomorrow"

A critical foundation for any maintenance planner lies in how equipment is organised and named in the CMMS. While it might sound like a purely administrative task, how you structure your assets and capture their metadata is pivotal for reliability, cost control, and efficient maintenance execution.

I've seen firsthand how a well-constructed equipment hierarchy can transform the entire maintenance operation and reporting structure, making it easier and clearer to understand asset relationships, analyse failures, and manage costs with confidence. Conversely, poorly structured data leads to confusion, inefficiency, and costly mistakes.

In this chapter we will explore how to create meaningful equipment structures, identify what asset metadata to gather, and deep dive into why these elements make all the difference in your role as a planner.

Structuring Equipment in a CMMS

When you first open a CMMS, you might be faced with hundreds or thousands of assets listed seemingly without order. This is where the equipment hierarchy comes into play - a way to organize assets so the system mirrors real-world relationships. Think of it as a family tree for your equipment, where every asset is connected to a parent or subsystem above it, and potentially children or components below.

At Amazon, detailed hierarchical models are in place which begin at the highest facility level, then drilled down to equipment operational groups/types, then to individual pieces of equipment, and finally to critical components. This allows users to see, for example, that a failed valve was part of a larger compressed air system, which was itself part of the utilities infrastructure supporting an entire warehouse. By understanding these relationships, the planner can anticipate ripple effects of equipment failure and plan maintenance to minimise downtime.

Typically, you'll find these hierarchy levels helpful:

- **Plant or Facility** - the highest level representing your site or warehouse.
- **Area or Zone** - subdividing the plant into manageable sections or production zones.
- **System or Line** - grouping assets that work together to perform a function.
- **Equipment or Asset** - the physical machines or devices.
- **Component or Part** - smaller elements like motors, bearings, or sensors.

This structure enables effective reporting and analysis at multiple levels. If a motor on a production line keeps failing, you can quickly drill down from the system level to the specific asset and even to the problematic component.

In the pharmaceutical industry, equipment hierarchy is often combined with criticality-based grouping - assets rated as 'critical' to production or safety receive different maintenance priorities. This dual approach gave planners a clearer focus and ensured resources were allocated where they mattered most.

Besides hierarchy, grouping assets by type or location is useful for certain reports or maintenance strategies. For example, grouping all pumps or motors across a facility allows trend analysis on that asset class, while grouping assets by floor or building aids technician routing and inspections. Total costs of maintenance within an Area/Zone/Asset Class can later all be rolled up to compare and draw actionable conclusions - e.g. why is conveyor line 1 costing us 3x that of line 2? Where are the top offending component costs coming from within that line, and why is that different across the 2 lines. Without this considered hierarchy in place, this cost rollup becomes very difficult and very manual, and the costs just sit as individual islands without any useful grouping.

Example Asset Hierarchy 1: Conveyor Line

Let's consider a typical conveyor line in a distribution centre or manufacturing plant. The hierarchy should reflect the physical and functional breakdown of the system so planners and technicians can easily locate and understand each component. The structure might look something like this:

Plant / Site
→ **Zone/Area**: Packaging Area
  → **System**: Primary Conveyor Line #3
    → **Equipment**: Conveyor Belt Frame
      → **Component**: Drive Motor
      → **Component**: Gearbox
      → **Component**: Belt Rollers
      → **Component**: Conveyor Belt
      → **Component**: Control Panel
      → **Component**: Sensors

In this structure, the planner can see the conveyor as a whole system but also drill down to specific parts that frequently require maintenance. This structure helps target inspections, track component lifecycles, and schedule replacements efficiently.

Example Asset Hierarchy 2: HVAC System

Utilities like HVAC are often critical for production environments and comfort. This type of hierarchy would be set up a little differently to the first example, but follows the same logic of grouping and component breakdown.

Plant / Site
→ **Zone/Area**: Office Building – Floor 2
→ **System**: HVAC System – Air Handling Unit (AHU) #5
→ **Equipment**: AHU Main Unit
→ **Component**: Fan Motor
→ **Component**: Heating Coil
→ **Component**: Cooling Coil
→ **Component**: Filters
→ **Component**: Dampers
→ **Component**: Control Valves
→ **Component**: Thermostats and Sensors

Here, metadata can include air flow rates, filter change schedules, and temperature setpoints, all essential for predictive maintenance planning.

Equipment Metadata: What to Capture and Why It Matters

Once you have your equipment structured, the next essential step is capturing comprehensive metadata for each asset. Metadata is the descriptive information that tells you everything about an asset beyond just its name. It provides context and supports daily decision-making.

According to ISO50001 standards on Asset Information, there are 6 criteria to define and measure data quality:

1. **Completeness** - is all required meta data present, or are there gaps which need attention?
2. **Consistency** - is the data type/quality consistent across all assets? Is the naming convention standardised so all similar asset types are called the same thing?
3. **Timeliness** - Is all data present with the latest reality?

4. **Uniqueness** - are all assets uniquely identified - i.e. no duplicates? They need to be unique in order to be properly searchable and identifiable.
5. **Accuracy** - is all the information linked to the assets an accurate and true reflection?
6. **Validity** - has the asset meta data been validated to know it is correct?

With those six asset data quality standards in mind, we can explore the key asset fields we should consider populating for your asset structure. Some metadata fields are non-negotiable, while others can be tailored to your organisation's needs:

- **Unique Identifier**: A unique asset tag or ID number is fundamental to prevent confusion-imagine trying to order parts for "Pump 2" when there are five pumps with that name.
- **Asset Name and Description**: Clear, standardised names help everyone, from planners to technicians, quickly identify the correct asset. This is usually a free text field, so relies on intentional input to an agreed naming convention.
- **Manufacturer, Model, and Serial Number**: These details are crucial for sourcing parts, accessing manuals, and warranty tracking. In many cases, the maintenance strategy would be tightly linked to the Manufacturer and Model numbers as an overall asset class.
- **Location**: Knowing exactly where an asset is physically located reduces technician search times and aids audits. This one varies by organisation, whether you have a lot of moving items which need locating, or quite static set up where the name identifies its location.
- **Installation and Commissioning Dates**: These dates feed into lifecycle management, helping forecast replacements or major refurbishments, and possibly manage any warranty cover.

- **Criticality Rating**: This guides later prioritisation of works knowing which assets could cause significant production loss or safety risks.
- **Maintenance Strategy**: Information about the assigned maintenance approach - be it preventive, predictive, or run-to-failure.
- **Technical Documentation Links**: Easy access to manuals and drawings saves time and ensures the right procedures are followed. This is a nice to have, but certainly effective for all users.
- **Failure Modes**: Documenting common ways assets fail supports targeted inspections and root cause analysis.
- **Spare Parts List**: Linking parts to equipment avoids delays in repair due to unavailable components. This also later helps roll up total installed quantities of all parts within a facility, which can help guide your spare part stocking strategy.
- **Operational Parameters**: Details such as pressure, temperature, or runtime parameters assist condition monitoring.
- **Cost Codes**: Tying maintenance activities back to budgets and accounting systems enables financial tracking.

The real value in asset metadata is its ability to transform raw data into actionable insight. Asset metadata deployed well in an organisation can be utilised not only to plan work but also to analyze failure trends, optimise spare parts inventory, and allocate maintenance resources efficiently. Without solid metadata, you're navigating in the dark.

The Crucial Role of Data Quality and Consistency

Metadata is only as useful as its accuracy and consistency. Inconsistent or missing data creates confusion, leads to misdirected work orders, and undermines reporting reliability.

As mentioned earlier, ISO55001 - the global standard for asset management - emphasises the importance of maintaining accurate, complete, and consistent asset information. This is not just about compliance; it's about giving your team the tools they need to succeed. ISO55001 requires that asset data supports decision-making, is accessible to those who need it, and is maintained to a high standard.

Metadata Audits

A critical component to reliable asset data is the deployment of audit programs where planners regularly review asset data and flag inconsistencies. This audit-feedback loop is critical for continuous improvement, and planners should play a leading role in driving up data quality standards.

Here is a possible checklist structure to help guide these audits:

<u>Identification & Naming</u>

- Each asset has a unique identifier (asset ID)
- Names follow a consistent, standardized naming convention
- Duplicate asset records have been removed

<u>Hierarchy & Relationships</u>

- Asset hierarchy reflects the actual physical and functional structure
- Parent-child relationships are clearly defined and accurate
- Assets grouped logically by location, type, or system

<u>Critical Data Fields</u>

- Manufacturer, model, and serial numbers are entered correctly
- Installation and commissioning dates are recorded

- Criticality ratings are assigned and updated as needed
- Maintenance strategy or task plans linked to assets

Location & Documentation

- Location data is precise and easy to understand
- Technical manuals, drawings, and service histories are linked
- Spare parts linked to relevant assets with part numbers

Data Consistency

- Data fields use controlled vocabularies or dropdown menus where possible
- No missing mandatory fields in asset records
- Metadata updated promptly after asset changes or maintenance

Cost & Operational Data

- Cost codes assigned for financial tracking
- Operational parameters (run time, temperature, pressure) entered where applicable

At many facilities, the scale of operations means that even small errors in asset data multiply quickly, causing scheduling conflicts and delays. Good practice is to establish quarterly metadata audits, where planners take ownership of reviewing certain asset groups. These audits, combined with feedback loops to the data entry and field teams, create a culture of accountability and continuous improvement.

Typically, planners who actively engage in these audits develop a deeper understanding of their equipment and maintenance priorities, which translates directly into better planning outcomes. It's not just about data hygiene - it's about empowering yourself and your team with reliable information.

Cost Rollup and Financial Transparency

One of the often-overlooked benefits of a well-built equipment hierarchy and rich metadata is the ability to roll up operational and maintenance costs effectively. When work orders are tied correctly to assets in the hierarchy, cost data flows upward - from individual components to entire systems and sites.

This capability is a game-changer for budgeting, forecasting, and investment decisions. For instance, planners can identify that a particular pump has a high total cost of ownership, factoring in repairs, parts, and labour, which may justify a replacement or a change in maintenance strategy.

Moreover, cost accountability is enhanced because departments or teams can see exactly where their budgets are spent, promoting ownership and cost control.

In your role as a maintenance planner, you are the steward of your organisation's asset information. A robust equipment structure and reliable metadata empower you to plan maintenance work effectively, analyse performance, control costs, and support compliance. This seemingly "behind-the-scenes" work ultimately drives the success of your maintenance program and the reliability of the entire operation.

From my hands-on experience in demanding environments like Amazon distribution facilities and global pharmaceutical plants, I can attest that investing time and effort in mastering equipment structures and metadata pays dividends many times over. It is not just data management - it's the foundation for operational excellence.

Key Takeaways

- Design your asset hierarchy thoughtfully, reflecting real-world equipment relationships but avoiding unnecessary complexity.
- Use standardised naming conventions and unique IDs to maintain clarity.
- Capture essential metadata for every asset and keep it current.
- Engage with your CMMS administrators to enforce data validation rules and mandatory fields.
- Schedule regular audits of equipment data and lead feedback sessions to continuously improve quality.
- Align your data practices with ISO55001 guidelines to support compliance and best practices.
- Leverage your asset structure and metadata to improve cost tracking, reporting, and decision-making.
- Train technicians and users on the importance of accurate asset data - they are critical partners in maintaining data quality.

Chapter 6

Building Job Plans That Work

"A technician should never have to guess what you meant"

In a reliability-centred maintenance program, the job plan is your frontline tool. It's what stands between reactive chaos and disciplined, repeatable execution. A great job plan saves time, reduces errors, and builds technician confidence. A poor job plan, on the other hand, is often worse than none at all - it creates ambiguity, breeds shortcuts, and undermines your CMMS processes.

As a maintenance planner, you're not just filling out forms - you're designing an experience for the technician. You should be seeking to lay out clear instructions, assemble tools and materials, and thinking through every step to minimise risk and inefficiency. When job plans are done well, you start to see the magic: work gets done right the first time, downtime drops, and planning becomes a strategic asset to the business.

What Are Job Plans and How Are They Used in a CMMS?

A **Job Plan** (sometimes called a task plan, maintenance instruction, or work template) is a predefined set of instructions and resources needed to complete a recurring maintenance activity. In most modern CMMS platforms, job plans are attached to:

- Preventive Maintenance (PM) tasks
- Predictive or condition-based tasks
- Corrective Work Orders for known, recurring issues (some sort of Standard WO functionality)
- Safety or compliance-related inspections

Each job plan usually contains some key elements:

- **Step-by-step instructions** - clear and concise, not too wordy, and in a logically flowing order

- **Required tools and parts** - help the techs prepare his toolkit and parts in advance, so they can arrive at the job with everything they require
- **Estimated labour and duration** - an estimate of how much time is required and how many techs, for shift leader resource allocation
- **Safety and permit requirements** - what safety precautions need to be considered and actioned before the job can start
- **Special notes or precautions** - other useful information/notes directed to the tech executing the work.

When a planner attaches a job plan to a scheduled work order in the CMMS, the system can automatically populate the relevant fields - saving time and ensuring consistency across teams and shifts. Over time, a library of quality job plans becomes one of the most valuable assets of the maintenance department.

Creating Effective Job Steps, Parts Lists, and Labour Estimates

Effective job plans don't need to be long, but they must be *clear*, *complete*, and *practical*. The aim here is clear communication from the planner/CMMS to the field techs, with ambiguity removed

Writing Clear Job Steps

Your job steps should guide the technician logically and safely through the task. Avoid generic phrases like "check motor" or "inspect pump." Instead, be specific:

Poor Example:
– Inspect bearings for wear

Better Example:
– Inspect motor bearings for signs of overheating, vibration, or

noise. Use stethoscope to confirm condition. Record observations in CMMS.

As a best practice to keep in mind, job plans are most effective when each step includes a **verb**, **target**, and **method** - that's what drives clarity and repeatability.

Including Parts and Tools

- **Parts**: List all parts required, including item numbers, quantities, and storage locations. Use CMMS linkages so planners can generate pick lists or ensure spares are stocked.
- **Tools**: Specify *exact* tools, especially if specialty equipment is needed (e.g., torque wrench, Megger tester). This helps technicians avoid wasted trips and improves readiness.

Estimating Labour

Estimating accurate time and labour resources is part of the planner's core value. Too short, and the job is rushed or delayed. Too long, and labour is underutilized. In time, actual feedback from completed work orders should inform your estimates. A good CMMS will allow side-by-side comparison between estimated and actual hours - this is gold for continuous improvement.

Designing Job Plans with the End User in Mind

This is where many planners get it wrong. A job plan that looks great in an office doesn't always translate to the field.

Technicians need job plans to be:

- Easy to follow
- Visually clear (e.g. bullet points, white space, diagrams)
- Available on the job (printed or digital on mobile/tablet)

- Specific to the asset and conditions they're facing

Involving your technicians in job plan development is one of the smartest moves you can make. Aim to regularly hold reviews with senior techs to refine your job plans. Their feedback improves clarity, eliminates unnecessary steps, and uncovers tribal knowledge that perhaps wasn't previously documented.

After implementing a new job plan, walk the job with a technician. Observe how they interpret and follow it. You'll often spot gaps that aren't obvious from behind a desk.

Checklists, SOPs, and Structured Data Fields

Job plans aren't just about instructions-they're also a vehicle for capturing consistent **data** from the field. Here's a breakdown of structured elements to consider:

Checklists

Use checkboxes for:

- Inspection tasks (e.g., "Check oil level")
- Visual confirmations ("No signs of leakage?")
- Repeatable steps that don't require detailed input

SOPs (Standard Operating Procedures)

SOPs should be referenced or embedded when tasks involve regulatory, safety, or detailed operational procedures. Ensure these are version-controlled and reviewed periodically.

Quantitative Data Input

Encourage measurement-based feedback where useful:

- Vibration levels
- Temperature readings
- Pressure values
- Wear tolerances

Structured fields in the CMMS should be used wherever possible-this enables trend analysis and predictive insights.

Yes/No and Conditional Fields

Useful for quick validations:

- "Did technician use Lockout/Tagout?"
- "Were replacement parts installed?"

These binary fields help automate compliance tracking and trigger escalation workflows when needed.

How to Avoid Technicians Going on Autopilot

This is a common and serious challenge - when job plans are too generic, too long, or too familiar, technicians can tend to start ignoring them. This leads to shortcuts, missed steps, and, eventually, preventable failures.

Tactics to Prevent Auto-Pilot Behaviour:

- **Keep it concise and relevant**: Don't include fluff or unrelated steps just to fill a template.
- **Update plans regularly**: If job plans never change, they're more likely to be ignored.
- **Add variation**: Include randomised checks or rotate inspection focus to keep attention sharp.
- **Use digital prompts**: If using a mobile CMMS, consider gating work orders so techs must input data before moving on.

- **Give feedback loops**: Share audit results or failure analysis that ties back to missed job plan steps.

Finally, remember that technician behaviour often reflects planner behaviour. If planners treat job plans as “set and forget,” so will the techs. But when planners visibly improve job plans based on feedback and performance, that mindset filters down to the field.

Key Takeaways

- Job plans are the frontline of reliability-when written well, they reduce ambiguity, save time, and boost technician confidence.
- Every job plan should include: clear instructions, accurate time and labour estimates, parts/tool lists, safety requirements, and relevant documentation.
- Write job steps that are specific and actionable-use verbs, targets, and methods to avoid vague directions.
- Involve technicians in job plan development and review-field feedback is essential for building useful, real-world instructions.
- Use job plans to collect structured data-checklists, measurements, failure confirmations, and yes/no validations feed back into your CMMS for better insight and tracking.
- Keep plans visually clear, concise, and accessible-especially when delivered digitally or in the field.
- Watch for “autopilot” behaviour-avoid it with concise steps, regular updates, and visible feedback loops.
- Treat job plans as living documents-every completed task is an opportunity to refine and improve.
- The best planners aren’t just process writers-they’re experience designers. Build your job plans with technician success in mind, and everything else improves with it.

Chapter 7

Preventive and Predictive Maintenance

"Prevention is cheaper than repair. Prediction is smarter than reaction"

Preventive and Predictive Maintenance (PM and PdM) are at the heart of any reliability-centred maintenance (RCM) strategy. As a Maintenance Planner, your ability to develop, refine, and optimise these approaches can make the difference between a reactive, fire-fighting maintenance culture and a smooth, efficient, and cost-effective operation. This chapter will explore in detail how planners can contribute to the structure, execution, and continuous improvement of preventive and predictive maintenance, and how modern tools and techniques, especially within a CMMS, support this journey.

Structuring and Improving PMs

Preventive maintenance refers to scheduled activities aimed at preventing equipment failures, extending asset life, and ensuring safety and compliance. It is structured to anticipate and reduce the likelihood of failures before they happen. However, not all PMs are equal. A world-class PM program is structured with intent, and a deep understanding of asset behaviour.

A good PM program includes:

- **Asset-specific instructions:** Generic PMs are a recipe for inefficiency. Each PM task should be tailored to the equipment, including context-specific instructions and safety precautions.
- **Failure mode targeting:** Maintenance tasks must be aligned with actual failure modes. For example, greasing a sealed bearing that doesn't require it not only wastes time but may lead to early failure.
- **Technician feedback loops:** The best insights often come from the people doing the work. A mechanism for capturing technician suggestions and observations ensures PMs evolve effectively.

As a planner, you're in the best position to coordinate these inputs. You sit at the intersection of operations, engineering, reliability, and maintenance. Use this to your advantage by driving structured PM reviews regularly.

Creating Job Plans for PMs

We explored in the previous chapter how every PM task should have an associated job plan in the CMMS. This job plan drives quality, consistency, and accountability.

Just to touch on some elements of a strong PM job plan again, in the context of improving your preventive maintenance regimes:

- **Clear job steps**: Each step should state what to do, how to do it, and any required measurements or indicators.
- **Parts and materials list**: This ensures everything is on hand and reduces delays.
- **Special tools or equipment**: Call out torque wrenches, IR cameras, or alignment jigs as needed.
- **Estimated labour time**: Supports scheduling and capacity planning.
- **Safety considerations**: Include lockout/tagout references or PPE.

The end goal is to facilitate a seamless execution. Technicians shouldn't have to guess, and supervisors shouldn't have to follow up. The plan is the blueprint.

Optimising PM Programs

Over time, PMs can become bloated, irrelevant, or duplicated. Optimisation is the ongoing process of refining your PMs so that every task delivers value.

Methods for PM optimisation:

- Failure mode and effect analysis (FMEA)
- Pareto analysis of downtime
- Technician field audits
- Historical performance reviews

Optimisation is not a one-time event. Schedule PM reviews quarterly or semi-annually to ensure continued alignment with operations and performance data.

Determining PM Frequency

Setting the right frequency for PMs is both an art and a science. Too frequent and you're wasting resources. Too infrequent and you risk failure.

Factors to consider:

- Asset criticality: Higher risk = higher frequency
- Historical failure trends
- Environmental conditions: Dust, moisture, vibration
- Manufacturer guidelines
- Available downtime windows
- Condition-based indicator
- Normal asset usage / run hours: Higher use = more maintenance

For example, a conveyor belt in a high-dust fulfilment centre may need weekly inspections, while a backup generator may only need quarterly checks if it's rarely used.

As a planner, be sure to collaborate closely with reliability engineers and field techs to determine appropriate intervals. Use the CMMS to track and adjust frequencies based on emerging data, and ensure that you maintain the mindset that frequencies are a dynamic thing. There is usually no need to be rigid about how often something is done, you can implement a starting point and adjust over time based on feedback and data.

RCM: Reliability-Centred Maintenance

RCM is a structured process that determines what must be done to ensure assets continue performing their intended functions. It involves identifying failure modes and selecting the most appropriate maintenance strategy.

RCM analysis includes:

- **Function**: What is the asset supposed to do?
- **Functional failure**: How does it fail?
- **Failure modes**: What causes the failure?
- **Effects**: What happens when it fails?
- **Consequences**: What's the impact?
- **Task selection**: What can be done to prevent or mitigate this?

This approach prevents "just in case" maintenance and ensures each task is rooted in real value. As a planner, you might not lead the RCM analysis, but you're critical in converting its outputs into job plans, schedules, and CMMS workflows.

Condition-Based Monitoring (CBM)

CBM involves using real-time data to monitor equipment health and predict failures. It reduces reliance on calendar-based PMs and instead triggers maintenance based on actual asset condition.

CBM tools and techniques:

- **Ultrasound analysis**: Early detection of steam trap or bearing issues
- **Infrared thermography**: Detecting heat buildup in motors or panels
- **Vibration analysis**: Identifying imbalance or misalignment
- **Oil analysis**: Checking for wear particles, water ingress, or contamination
- **Electrical testing**: Motor current signature analysis, insulation resistance

Planners will often support a Condition Based Monitoring programme by:

- Building CBM inspection routes within the CMMS
- Scheduling CBM inspections and follow-ups
- Ensuring CBM data is logged and tracked in the CMMS
- Initiating work orders based on CBM triggers (unless auto-triggered)
- Developing job plans that include CBM-specific tasks

How CBM Supports RCM

RCM sets the strategy-CBM enables the tactics. For each identified failure mode in RCM, CBM can offer a way to detect degradation before it becomes a failure.

Rather than blindly replacing motor bearings every 12 months for example, a CBM approach using ultrasound or vibration sensors may detect wear after 18 months - or in some cases, not at all. That saves cost, reduces waste, and improves uptime.

The planner's role is to make CBM part of everyday maintenance. Ensure sensors are read, data is analysed, and actionable tasks are generated and closed through the CMMS.

Avoiding Under- or Over-Maintenance

Too little maintenance and you suffer unplanned failures. Too much, and you increase cost, introduce new failure risks, and overload your workforce.

Use a combination of the following tools to maintain that balance correctly:

- RCM analysis
- Failure trends and asset history
- Technician feedback
- Work order audit reports
- CMMS analytics dashboards

Critically here, as a planner, you need to actively challenge assumptions. If a maintenance task has been done monthly for three years but never prevented a failure risk, ask why. Can it be removed/reduced? Replaced with a PdM method? Adjusted?

Leveraging CMMS for PM & PdM

Modern CMMS platforms allow you to automate and track all aspects of PM and PdM:

- Scheduled task generation
- Job plan standardisation
- Parts and labour forecasting

- CBM data integration
- Technician performance tracking
- Cost analysis by asset

Invest in your CMMS skills. Learn to use templates, workflows, and auto-trigger rules. Partner with IT or system admins to build custom fields or reports if necessary. Your CMMS is the nerve centre of your maintenance strategy.

Preventive and predictive maintenance strategies are central to improving reliability and reducing downtime. As a planner, you are the orchestrator that brings RCM theory and CBM technology into practical daily execution.

Key Takeaways

- PM and PdM are core to reliability-centred maintenance. As a planner, your influence lies in how well you structure, optimise, and embed these strategies into everyday maintenance.
- Strong PM job plans drive execution quality. Clear instructions, defined tools and parts, realistic labour estimates, and safety steps are essential for consistency and performance.
- PMs must evolve. Use field feedback, failure history, and performance data to refine tasks, remove waste, and adjust frequencies. Stale PMs are as dangerous as missing ones.
- RCM provides the strategy-planners deliver the execution. Even if you don't lead RCM analysis, you're critical in converting outcomes into effective job plans, schedules, and CMMS processes.
- Condition-Based Monitoring (CBM) improves precision. It enables you to move beyond calendar-based PMs by responding to real asset health indicators.
- Planners make CBM practical. Your role is to ensure data is collected, reviewed, and turned into timely, executable work.
- Avoid under- and over-maintenance. Question default frequencies, validate with data, and pursue smarter alternatives where possible.
- CMMS is your most powerful tool. Use it to automate schedules, integrate CBM data, track effectiveness, and inform future improvements.
- Your work reduces downtime, saves cost, and protects technicians. PM and PdM aren't just tasks-they are your levers for long-term reliability success.

Chapter 8

Work Order Management

"Your work orders tell the story of your whole maintenance system"

Effective work order management is the beating heart of a well-functioning maintenance planning process. It is where the planning work transitions into execution, and where the CMMS becomes a dynamic tool, not just a passive database. Work Orders (WOs) represent the structured communication and tracking of maintenance work. When managed well, they enable clarity, accountability, and continuous improvement.

In Chapter 3, we discussed the lifecycle of maintenance work and how planners contribute to each step. Work order management touches nearly every stage of that lifecycle, but this chapter focuses specifically on how to manage the work order process comprehensively, from intake to closure.

Intake: Where Work Orders Begin

Work orders can originate from a variety of sources. Understanding and managing these inputs is essential for prioritizing effectively and ensuring good planning outcomes.

Sources of Work Orders:

- **Planned Maintenance (PM):** These are routine jobs generated at regular intervals based on time or usage.
- **Predictive Maintenance (PdM) and Condition-Based Monitoring (CBM):** Triggered based on condition thresholds or anomalies detected via technologies such as vibration analysis, thermography, or oil sampling.
- **Reactive Maintenance:** Unplanned requests typically initiated by operations or technicians in response to breakdowns or abnormalities.
- **Customer Ticket Requests:** Submitted by internal stakeholders such as production, quality, or safety teams.
- **Project Work Orders:** Related to ongoing or one-off projects, upgrades, or installations.

- **Follow-Up Work Orders:** Arising from findings during other maintenance activities.

Regardless of source, all work orders should be captured within the CMMS to allow for structured assessment, tracking, and continuous improvement.

Planner's Role: The planner should be the first filter. Before any work order is acted upon, it should be reviewed for clarity, duplication, and alignment with existing plans.

Tips for Managing WO Intake:

- Create templates or forms that ensure minimum required information is provided when a WO is submitted, especially when coming from external stakeholders (e.g. Operations)
- Continuously coach other teams on the importance of detailed work request submissions.
- Periodically review incoming WOs to identify trends in unclear or poor-quality submissions and provide feedback.

Scope: Defining What the Work Order Should Cover

One of the most critical functions a planner performs is ensuring that each work order is clearly scoped. Without a defined scope, execution becomes chaotic, outcomes are inconsistent, and post-job analysis becomes difficult.

Clarifying the Request

Work orders initiated by others may be missing key details. The planner or someone in a CMMS Administrator role must bridge that gap between request and execution.

Common Issues:

- Vague problem descriptions ("It's noisy")

- No location or asset ID specified
- No urgency or business impact explained

Planner Interventions:

- Contact the requestor to clarify the issue
- Use site knowledge or previous work history to inform the scope
- Add additional details or preliminary diagnostics based on available data

Controlling Scope Creep

Scope creep can derail even simple jobs. Additional tasks discovered during execution should be logged separately as follow-up work orders unless they are essential to safe completion.

Good Practices:

- Define boundaries in the job plan (e.g., "Inspect and replace bearings on Motor 2 only.")
- Coach technicians to raise follow-up WOs for additional findings
- Use CMMS workflow controls where available to enforce review of follow-up work before scheduling

Estimating Resources

A key planning responsibility is estimating what's required:

- **Time**: How long will it take? Include travel/setup/cleanup/CMMS admin
- **Skills**: Is specialist training or certification required?
- **Tools**: Are special tools needed?
- **Parts**: Are spares available? Will they need to be ordered? Can a job in a box be prepared in advance?

Estimates should be recorded in the CMMS. Over time, actuals vs. estimates can be used to improve planning accuracy, as long as the data quality input by the field team is accuracy and consistent.

Prioritisation: Balancing the Workload

Not all work is created equal. Prioritisation ensures the right work is done at the right time. It is not only about what seems urgent, but what is most important to system reliability, safety, and operational continuity.

Planner's Role in Prioritisation: While planners don't always assign the final priority, they provide the structure and discipline to make priority decisions effectively.

Key Considerations:

- **Safety risk**: Is there a risk of injury or non-compliance?
- **Production impact**: Will this lead to downtime or quality loss?
- **Asset criticality**: Is the asset essential or redundant?
- **Cost implications**: Will delay lead to further damage or cost escalation?

Dealing with Priority Pressure

It is common for requestors to assign high priorities to all their work orders. Planners must resist this trend diplomatically.

Advice:

- Educate stakeholders on the priority matrix or model used by the maintenance team
- Offer visibility into backlog and current work queue to justify scheduling decisions
- Use CMMS features to auto-categorise based on asset and issue type

Closure: Ensuring Work is Complete and Learnings Are Captured

Work order closure is often the most neglected stage in the lifecycle. Yet it is where the value of good planning, execution, and data converge.

What Should Be Recorded?

The CMMS should capture the following as part of a comprehensive closure:

- **Labour time**: Who did the work and how long it took
- **Parts used**: Including part numbers and quantities
- **Failure codes**: What was the failure mode or cause?
- **Closing comments**: Describe what was found and what work was done
- **Date/time of completion:** Often taken by the system date/time at point of WO closure
- **Any follow-up actions needed**

Planner Oversight

Not every work order needs a post-closure review, but strategic sampling or automated audits can maintain quality. For critical equipment or failures, review should be mandatory.

Review Triggers:

- High-cost jobs
- Unexpected downtime
- Repeated failures on the same asset
- Technicians flagging a follow-up

Improving the Closure Process

- Train technicians on how to enter good close-out notes

- Use dropdowns where possible (e.g. failure codes) to reduce variability and ensure validation
- Provide examples of good vs. poor closeout comments as part of training materials
- Perform regular audits and provide feedback

Managing Work Order Backlogs

A common challenge in many maintenance teams is the ballooning backlog of open work orders. While a backlog is not inherently bad (some low-priority work can and should wait), unmanaged backlogs lead to inefficiency and risk.

Planner Strategies for Backlog Control:

- **Backlog Review Meetings**: Hold regular sessions to review aging work orders
- **Age Triggers**: Use CMMS to flag WOs older than a certain threshold (e.g. 90 days)
- **Auto-Close Rules**: For certain types of low-priority work, consider auto-closing after a period with supervisor sign-off
- **Categorisation**: Use work order types and priority codes to sort backlog intelligently

Balancing Planned and Reactive Work: A healthy maintenance program aims for a high ratio of planned vs. reactive work. Work order management is central to this goal. If a large percentage of work orders are being created as urgent reactive tasks, investigate the root cause:

- Are PMs missing key failure points?
- Is CBM data being ignored?
- Are jobs being delayed or canceled due to parts/tools/people not being ready?

Work order management is more than just a mechanical process. For planners, it is a high-leverage activity that connects the tactical and strategic dimensions of maintenance. The intake process defines the challenge; the scope frames the solution; prioritisation drives efficiency; and closure captures the lessons that power long-term improvement.

Used wisely, the work order process becomes a feedback engine. Each job feeds the next generation of better planning, and the planner is at the centre of that learning loop.

When structured and supported by a capable CMMS, and with disciplined planner involvement, the work order process becomes a key driver of equipment reliability, technician effectiveness, and organisational confidence in the maintenance function.

Key Takeaways

- Work orders are the core operational unit of maintenance-when managed well, they drive clarity, accountability, and long-term improvement.
- Intake is critical-planners must filter and validate requests from multiple sources, ensuring each has enough detail to proceed and fits into the overall workload.
- Scope definition ensures each job has clear boundaries, accurate expectations, and avoids scope creep. Vague requests lead to inefficiency and inconsistent results.
- Estimate the right resources, tools, time, and skills needed for each job. Track estimated vs actuals over time to refine planning accuracy.
- Prioritisation must be structured-planners should guide decisions based on asset criticality, risk, and business impact, not who shouts loudest.
- Use CMMS features to reduce pressure-driven prioritisation and introduce visibility into backlog queues for fairer scheduling decisions.
- Planners should audit closure data-especially on critical jobs or repeated failures-to drive improvement and maintain CMMS data integrity.
- Backlog health matters-manage aging work orders through review cycles, categorisation, and intelligent auto-close rules for non-critical tasks.
- A disciplined work order process creates a feedback engine-each job feeds better data into the next, with the planner steering that learning loop.
- With strong CMMS support and clear planner ownership, work orders become more than tasks-they become the foundation of equipment reliability.

Chapter 9

Spare Parts and Materials Readiness

"No plan is complete until the right parts are in the right place, at the right time"

Spare parts management is a critical, yet often underappreciated, component of maintenance planning. While the technical skills of creating job plans and work orders receive a lot of attention, the material aspect; specifically managing spare parts, can be the difference between a smooth maintenance event and a costly delay.

For planners, understanding and managing the materials side of maintenance is not about becoming a procurement specialist or warehouse operator. It is about ensuring that the right part is available at the right time to support reliable, efficient execution of work. In different organisations the planner role will have different levels of responsibility within the spare parts management arena, but whatever your responsibility is now, this is a topic well worth having a thorough understanding of.

Why Spare Parts Matter in Maintenance Planning

No matter how detailed or well-written a job plan is, if the required parts are not available, the job will not be completed on time. This delays work, reduces planner credibility, increases reactive work, and undermines the very essence of proactive maintenance.

Spare parts influence planning in several ways:

- **Lead time:** Some critical parts have long procurement lead times, and without forward planning, this can impact schedules significantly.
- **Inventory cost:** Excess stock ties up capital, while stockouts increase risk.
- **Job readiness:** Planning must include checking that parts are in stock or initiating the purchase process early.
- **Work deferrals:** Without parts, jobs may be pushed out or rescheduled, potentially missing PM intervals or reliability-critical windows.

- **Part Criticality:** Some parts have the capacity to bring down an entire system, whereas some are less impactful. This criticality will affect the stock holding strategy.

As a planner, you sit at the intersection of maintenance execution and materials readiness. Your role is to align both worlds.

The Planner's Role in Spare Parts Management

While procurement (and sometimes dedicated stores teams) may own the transactional and logistical elements of spare parts management, the maintenance planner should play a strong coordinating role in developing the strategy. Here are key planner responsibilities:

1. Identifying Parts Required for Job Plans

Every job plan should include a clearly defined list of required parts, with accurate part numbers and quantities. This ensures that stores or procurement can respond efficiently.

- Use previous work orders and technician feedback to refine parts lists.
- Cross-reference parts with the CMMS to ensure correct identification.
- Flag any obsolete or superseded parts.

2. Verifying Availability

When planning a job, verify:

- If the required parts are in stock
- The stores location of parts is accurate and clear (bin location)
- If special tools or one-time-use items are available

If this is managed by another dedicated team, ideally planners will use the CMMS integration with inventory systems or communicate with the stores team directly to confirm availability.

3. Reserving or Kitting Parts

If the CMMS or ERP allows, planners can place reservations on required parts or request a kit to be prepared.

- Kitting simplifies job execution, minimizes search time, and reduces errors.
- Kits can be tagged with work order numbers and held in staging areas.
- Reservations often consider the stock effectively 'consumed' so reordering can commence immediately to avoid future stock issues.

4. Collaborating with Procurement and Stores

Planners having a strong working relationship with procurement and inventory teams helps ensure that maintenance priorities are supported by materials readiness.

- Share forward-looking schedules so procurement can plan for future demand.
- Identify recurring parts usage and align with re-order thresholds.
- Escalate long-lead items early to avoid critical path delays.

Spare Parts Meta Data: What to Capture and Why

Just as we explored that assets have metadata, so too do spare parts. High-quality data enables better planning, procurement, and inventory control. Here are some key metadata fields to track against your parts:

- **Part Number** (unique identifier)
- **Description** (clear, standardized, searchable)
- **Associated Equipment or BOMs** (where the part is used)
- **Unit of Measure**
- **Preferred Vendor or Manufacturer**
- **Minimum/Maximum Stock Levels**
- **Lead Time**
- **Unit Cost / Value**
- **Criticality** (impact of not having the part)
- **Storage Location (Bin/Zone)**
- **Hazardous Material Status** (if applicable)

Metadata accuracy directly impacts the efficiency of parts searches, order accuracy, and job readiness. For planners, this metadata informs decisions such as:

- Is a part commonly used and should it be stocked?
- Should I schedule this job next week or wait until the part is in stock?
- Is there a risk of stocking a superseded or incorrect version of a part?
- What is the risk vs cost consequences of the stock levels?

Key Data Quality Considerations

As with asset metadata (Chapter 5), data quality for spare parts is essential. Planners should help enforce standards and report inconsistencies.

Common Data Quality Issues:

- Duplicate part numbers
- Inconsistent naming conventions
- Missing or incorrect BOMs
- Obsolete parts listed in job plans

- Empty or incomplete fields

Planner Actions:

- Work with technicians to validate parts during execution
- Provide feedback loops to stores when parts are misidentified
- Use CMMS reports to audit job plans for missing or outdated part info

Aligning Spare Parts with Job Planning and Reliability

Planning isn't just about putting jobs on a calendar-it's about ensuring all inputs to maintenance are aligned. That includes parts. Here's how spare parts management supports broader planning goals:

Reliability-Driven Planning

- Jobs on critical equipment must be fully kitted to avoid delays.
- Avoid substituting parts ad hoc, which can lead to compatibility issues or premature failures.
- Correct (like-for-like) parts are available to avoid start-up issues after replacement.

Work Order Readiness

- Planners should use a checklist during planning:
    - Are all parts identified?
    - Are they available or ordered?
    - Are parts in usable condition (e.g., calibration date, shelf life)?

CMMS-Integrated Planning

Many modern CMMS platforms allow planners to:

- Link parts directly to job plans and work orders
- View inventory levels in real time
- Trigger reorder notifications
- Auto-create material pick lists for kitting

Use these tools to streamline the planning workflow and reduce manual steps.

Spare Part Criticality and Stocking Strategy

Not all parts should be stocked. Inventory strategy should be guided by:

- **Asset criticality**
- **Part criticality**
- **Failure frequency**
- **Part lead time**
- **Quantity installed on the plant/site**
- **Cost and storage footprint**

Planners are key stakeholders in helping define which parts justify being stocked versus ordered as needed. Participate in critical spares reviews and recommend stocking changes based on observed work order history or known pain points in execution. At minimum, minimum and maximum stock levels should be assessed on an annual basis and updated as necessary.

Stock Control: Trust, but Verify

A well-planned job can still fail if the expected parts aren't actually available when needed. While CMMS or ERP systems may show certain stock levels, these numbers are only as good as the physical inventory behind them. That's why stock control is essential - planners must be confident that what's listed in the system exists on the shelf.

Stock inaccuracies can lead to:

- Last-minute job cancellations
- Emergency orders and unplanned downtime
- Loss of planner and system credibility

Good stock control practices include:

- **Weekly Spot Checks**
  Select a small number of high-usage or critical parts to verify physically. Over time, this builds confidence in your system data - or reveals where improvements are needed.
- **Cycle Counting**
  Work with the stores team to implement rotating inventory audits. Instead of a single large count, split the inventory into categories or zones and count a section each week/month.
- **Annual Full Inventory**
  A full inventory check once per year remains industry best practice. It helps correct deeper discrepancies and aligns physical stock with digital records.
- **Planner's Role**
  While the stores team typically owns the inventory, planners should:
    - Flag frequent mismatches between CMMS and physical availability
    - Track jobs that were delayed due to stock inaccuracies
    - Provide feedback on part movement trends to improve reorder points

Accurate stock levels aren't just a warehouse concern - they're a planner's foundation. By partnering with inventory teams to maintain stock accuracy, you ensure your planning remains executable, efficient, and trusted.

Budget Awareness and Cost-Conscious Parts Strategy

While the technical side of maintenance planning often takes centre stage, planners must not lose sight of financial stewardship. Effective spare parts management has a direct impact on maintenance budgets and broader organisational cost performance.

The Cost of Overstocking

Holding excess inventory ties up valuable capital. Items that sit idle on shelves for months or years represent opportunity costs and increase the risk of obsolescence, especially in rapidly evolving manufacturing environments. Unused parts may:

- Be written off as obsolete stock
- Exceed their shelf life (e.g., seals, lubricants)
- Be incompatible with upgraded equipment

A planner who considers consumption rates and forecasted maintenance needs can avoid over-specifying parts and help maintain leaner, more agile inventory levels.

Cost of Stockouts and Emergency Purchases

On the flip side, not stocking enough can result in urgent, expedited procurement requests that carry significant price premiums. The planner's role is to balance these risks:

- Identify recurring part usage to guide minimum stock levels
- Work with procurement on vendor agreements that allow just-in-time supply without stockouts

Collaborating on Budget Management

Planners are not budget owners, but they influence budget outcomes significantly. Working closely with the stores and finance teams, planners can:

- Flag high-value, low-usage parts for review
- Recommend bundling of orders to reduce shipping costs
- Support justification for critical spares based on failure history

Supporting a Strategic Spare Parts Policy

An effective parts strategy is guided by:

- Regular asset criticality reviews
- Cost-benefit analysis of stocking decisions
- Risk-based inventory planning

Planners should advocate for regular formalised reviews where engineering, procurement, and finance collaborate in order to:

- Define critical spares
- Eliminate obsolete stock
- Adjust reorder points based on usage trends

In this way, the planner becomes a contributor not just to operational readiness, but to sound financial decision-making.

Planners play an essential bridging role between maintenance and materials. Though they don't own the inventory, they are responsible for ensuring that the work they plan is executable with the parts and materials available.

When done well, spare parts management enhances job readiness, improves technician productivity, reduces delays, and supports a more reliable and cost-effective maintenance program.

In the next chapter, we'll explore the scheduling process-how to bring all the planned and prepared work together into an executable, efficient weekly maintenance schedule.

Key Takeaways

Spare parts are a critical enabler of planned maintenance-without the right part, even the best job plan will fail to deliver on time or at all.

- Planners must understand the relationship between materials and execution. You're not just planning the job, you're ensuring it's executable.
- Key planner responsibilities include:
    - Identifying and specifying required parts in job plans
    - Verifying availability and bin location
    - Requesting reservations or kits for upcoming work
    - Collaborating with stores and procurement on stock and supply timelines
- High-quality spare part metadata (e.g., part numbers, lead times, usage, vendor info) is essential for accuracy, efficiency, and cost control.
- Planners can use the CMMS to link parts to jobs, generate pick lists, and track inventory-minimising manual steps and delays.
- Data quality matters-planners should help eliminate duplicate parts, validate BOMs, and enforce naming conventions.
- Inventory strategy should be based on criticality, usage frequency, lead time, and cost impact-not gut feel or habit.
- Stock control is essential-partner with inventory teams to verify actual availability through cycle counts, spot checks, and full inventories. Accurate stock data underpins reliable job execution.
- Overstocks drain capital and risk obsolescence; stockouts cause costly delays. Planners must help balance these risks through informed planning.
- While not budget owners, planners influence spend by:
    - Reducing emergency orders
    - Helping set reorder levels
    - Advising on spare stocking policies

- Planners should support cross-functional reviews to refine spare parts strategy, eliminate obsolete stock, and drive risk-based stocking decisions.
- Spare parts aren't "someone else's problem"-they are integral to the planner's ability to deliver reliable, repeatable maintenance.

Chapter 10

Scheduling and Co-ordination

"Planning builds readiness - scheduling brings it to life"

Maintenance planning and scheduling are often bundled together as one function, but the two serve distinct roles. So what's the difference?

Planning defines *what* needs to be done and *how*, while scheduling determines *when* it will be done and *by whom*.

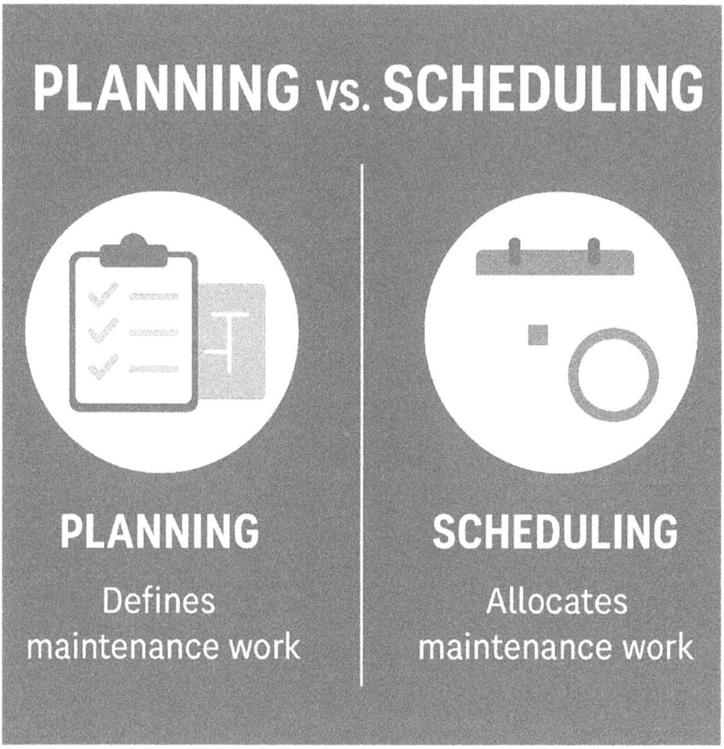

This chapter focuses on the scheduling and coordination piece - translating well-prepared work orders and job plans into executable action within the realities of an operational environment.

While planning ensures technical readiness, it is through effective scheduling that the maintenance strategy becomes tangible. Without a robust scheduling and coordination framework, even the most well-built plans can stall.

Long-Term Planning

The Yearly PM Plan

Creating a 12-month preventive maintenance (PM) schedule is a foundational scheduling task. This high-level plan maps out when each PM should occur across the year and ensures regulatory, OEM, and safety-critical inspections are never missed.

Key aspects of a robust yearly PM plan:

- **Even workload distribution**: Avoid front-loading or bunching tasks in certain months unless driven by shutdowns or operational downtime.
- **Coordination with production/operations**: Align PMs with planned outages, low-demand seasons, or shutdown windows to minimize disruption.
- **Resource planning**: Estimate the annual labour, parts, and contractor support required to complete the plan and use this to inform budget cycles.

As a planner, you should engage with stakeholders such as operations, safety, and procurement when developing the yearly plan. This collaborative approach builds alignment and sets expectations early.

Finding Efficiencies Through Bundling

A shrewd planner looks beyond the calendar and finds opportunities to group PMs in a way that makes execution more efficient. For example:

- Grouping multiple PMs on equipment within the same system or location (e.g. several conveyors on one line).
- Combining inspections that require the same skillset or tools.

- Coordinating tasks to reduce changeovers or setup time for technicians.

This “PM bundling” approach supports:

- Reduced maintenance downtime - less impact on your customer
- More efficient technician use.
- Consolidated parts/tooling preparation.

Don’t just group for the sake of grouping - ensure bundled tasks are logistically compatible and do not overwhelm the team on execution day.

The Weekly Planning Cycle

While the yearly PM plan provides a strategic view, the weekly planning cycle is where execution happens. It requires active coordination, detailed communication, and realistic scheduling based on ever-changing available resources.

Looking Ahead

The planner’s role is to look forward ideally 1 to 4 weeks ahead, and begin preparing the work packages and support elements necessary for successful execution. For work to be included in the weekly schedule, it should be:

- **Fully scoped**: All information clearly defined in the job plan.
- **Materially ready**: All required parts, tools and consumables either on hand or confirmed to arrive before the scheduled date.
- **Approved and prioritised**: Reviewed through a prioritisation meeting or weekly planning huddle.

By building this “ready-to-schedule” backlog, planners ensure the schedule isn't reactive but intentional and informed.

Resource Analysis

One of the most challenging aspects of scheduling is balancing the required work against available labour. This includes:

- Full-time technicians
- Contractors or temporary support
- Known absences (leave, training, sickness)

A good scheduler will:

- Forecast technician availability week-to-week.
- Track resource gaps due to unplanned absences.
- Work with team leaders to manage overtime, shift swaps, or deferrals when necessary.
- Amend the short-term schedule to avoid any overloaded shifts.

Tools such as technician capacity heatmaps, Excel-based resource loading charts, or CMMS-based scheduling views are incredibly useful here.

Depending on your organisation's conditions, aim to build a weekly plan that is 70-80% loaded. This leaves room for emergency work and unexpected shift absences without derailing the entire schedule.

Making an Achievable Plan

A plan that looks great on paper but fails in practice undermines trust in the planning function. The weekly schedule must be *realistic* - factoring in human limitations, shift lengths, travel time, and competing priorities.

What makes a schedule achievable?

- Accurate labour hour estimates in job plans.
- Awareness of what jobs are "must-do" versus "nice to have."
- Consultation with field leadership on what is reasonable.
- Any planned maintenance equipment downtime is fully agreed with operations and sufficient to complete the scheduled works.

Avoid the trap of overscheduling just to tick boxes. It's better to under-promise and over-deliver.

Assigning Work Orders in the CMMS

How and when work is assigned within your CMMS can affect technician clarity, accountability, and workload balance.

Best practice recommendation:

- Planners assign work to the correct shift/team, based on the timing and skill requirements.
- Team leaders or maintenance managers assign to specific individuals during their daily or shift-level meetings.

This approach ensures:

- Flexibility for local team leadership to adjust in real-time.
- Clear ownership and accountability within shifts.
- Reduced risk of bottlenecks from assigning to unavailable individuals.
- Works are aligned to any required specific skillsets within the maintenance team.

The CMMS should be your central scheduling tool. Use features like drag-and-drop scheduling boards, shift calendars, and colour-coding to visualise upcoming work and identify gaps. Depending on the capability of your specific CMMS, it maybe that third party software (e.g. Excel) is required to put the plan in a more visual format to analyse and tweak before updating in the live system.

Preparing the Team

Even the best schedule can fall flat if the team isn't informed and aligned. As a planner, part of your role is enabling the team to hit the ground running. This includes:

- **Publishing the schedule** at least 48 hours in advance of the work week.
- **Holding planning/scheduling meetings** with technicians and supervisors to walk through key tasks.
- **Distributing work packages**, including job plans, permits, drawings, or risk assessments.
- **Highlighting special tasks**, unusual risks, or work requiring cross-team support.

By preparing the team well in advance, you avoid delays, misunderstandings, and half-executed jobs.

Shutdown Scheduling: Bringing Order to Chaos

Large-scale shutdowns - whether they are department-level, full site outages, or major system overhauls - present unique challenges and opportunities for maintenance teams. These windows are often the only times certain tasks can be executed safely and thoroughly, especially on equipment that must run continuously during standard operations.

For the maintenance planner, a shutdown is not just a scheduling exercise - it's essentially a project. The scale, complexity, and volume of work involved demand a level of coordination that's fundamentally different from business-as-usual scheduling. Done well, the planner brings order to what can otherwise be chaos.

Making Maximum Use of the Planned Downtime

Shutdowns provide a valuable opportunity to:

- Perform invasive inspections or maintenance that can't be done during operations.
- Complete grouped tasks across systems or production lines for efficiency.
- Implement improvement projects (installing new equipment, upgrades, relocations).
- Conduct statutory inspections or external audits.

The planner's job is to ensure that every hour of downtime is used to maximum value, without overwhelming the teams or risking poor execution due to overloading.

The Planner's Role in Shutdowns:

The planner often becomes a *mini project manager* in these situations, coordinating not just tasks, but people, materials, timing, and contingency plans.

Key responsibilities include:

- **Work identification**:
    - Work orders that have been deferred due to operational restrictions.
    - Projects or CAPEX work that need downtime to implement.
    - Inspections required by regulation or policy (e.g. pressure vessel checks, fire system validation).
- **Work scoping and validation**:
    - Confirming the accuracy of job plans.

    - Ensuring drawings, permits, isolations, and tooling are available.
    - Vetting all tasks to avoid last-minute surprises.
- **Prioritisation and work selection**:
    - Every stakeholder will want their task included. The planner helps leads in filtering what is *must-do* vs *nice to do*.
    - Work should be risk-ranked and weighed against available time and labour.

A planner's ability to say "no" or "not this time" is just as important as their ability to say "yes." Being selective preserves schedule quality and execution integrity.

- **Resource alignment**:
    - Coordinating internal teams and external vendors.
    - Creating detailed resource loading charts to ensure coverage without conflict.
    - Accounting for support staff like safety observers, cleaners, or riggers.
- **Sequencing and time-based coordination**:
    - Carefully mapping which tasks must precede others (e.g. isolations before cleaning, inspection before overhaul).
    - Identifying critical path items and milestone checks.
    - Building in contingency time for overruns, rework, or bad weather.
- **Material kitting and logistics**:
    - Ensuring all parts are available, staged, and clearly labelled.
    - Reducing wasteful downtime looking for items or waiting on deliveries.
    - Ensuring stock transactions are accurately accounted for during an intense period of high part consumption.
- **Communication**:

  - Publishing the shutdown schedule well in advance.
  - Running daily planning meetings during execution.
  - Providing live updates to leadership on progress, blockers, and safety.

A well-run shutdown is defined not by the number of jobs completed, but by the number of problems you prevented through foresight and coordination.

Managing Complexity: Multiple Teams, Multiple Tasks

During a shutdown, dozens of people may be working in the same area - often with conflicting priorities. Safety, access, and sequencing all become critical.

The planner brings structure and clarity by:

- Creating visual Gantt charts or timeline boards, mapping out every task and crew.
- Coordinating isolation plans and lockout/tagout procedures to avoid overlap.
- Confirming that high-risk work (confined space, hot work) is scheduled during lower-activity periods where possible.
- Making site maps or task zones to prevent congestion and confusion.

Some sites use colour-coded badges or area permits to manage who is allowed to be where during a shutdown - planners should be involved in designing and enforcing this structure alongside safety teams.

Preparing for Restart: Time to Test and Verify

One of the most overlooked aspects of shutdown scheduling is the return to service phase. A common failure point is cramming too much work into the shutdown window, leaving no time for testing, inspections, or safe restarts.

The planner's role includes:

- Scheduling pre-start inspections, ensuring all components are correctly reinstalled and aligned.
- Allowing time for instrument calibration, controls testing, or software uploads.
- Reserving capacity for function checks, like test runs of motors, conveyors, or HVAC systems.
- Coordinating with production teams on ramp-up timing and buffer material needs.

Never assume that once the last bolt is tightened, the job is done. Time must be built in for unforeseen issues, and for the team to safely bring the system back online.

Here are some of the most valuable shutdown planning tips:

- **Start planning early**. Three to six months in advance is ideal for large-scale outages.
- **Keep a lessons learned document** after each shutdown. Review what worked, what didn't, and what needs to be done differently next time.
- **Have a cut-off point** for adding new work. Tasks added too late rarely succeed and often introduce risk.
- **Use a command centre model** during the shutdown. A single point of contact (usually the planner or project lead) should field updates and issue decisions centrally.

Daily Execution and Follow-Up: Keeping the Pulse on the Floor

One of the key responsibilities of a maintenance planner goes beyond creating the perfect schedule - it's about staying connected with the reality on the ground. The day-to-day execution phase is where plans meet real-world conditions, and the planner must stay agile, responsive, and communicative to keep maintenance flowing smoothly.

Staying Connected: Ears on the Ground

The best planners develop a strong, ongoing connection with the technicians and engineers executing the work. This "ears on the ground" approach helps them:

- Identify emerging issues early, before they escalate into bigger problems.
- Understand constraints or bottlenecks in real time.
- Gauge morale and workload, which informs future scheduling decisions.
- Spot opportunities to improve job plans based on feedback.

This requires more than just monitoring a CMMS screen - it means actively engaging with the field teams on site, walking the floor regularly, and building trust so that teams feel comfortable sharing challenges openly.

Daily Stand-Up Meetings: A Hub of Communication

Daily stand-ups are an effective tool for keeping everyone aligned. These short, focused meetings typically involve:

- Reviewing the status of planned work orders and any carryovers.
- Highlighting critical tasks or high-priority issues.

- Identifying resource shortages or equipment availability conflicts.
- Discussing safety alerts or permit updates.
- Planning adjustments for the day or the coming days.

These meetings foster transparency and collective problem-solving, empowering teams to adapt quickly when circumstances change. The planner often acts as the communication hub, ensuring issues are escalated promptly and resources are reallocated as needed.

Handling Rollover of Incomplete Work

Not all planned work gets completed on schedule. Delays can happen for many reasons: unexpected faults, lack of parts, absent personnel, or emergent higher priority jobs.

Managing rollover work thoughtfully is crucial to avoid creating backlogs that spiral out of control.

Key considerations include:

- **Assessing impact:** How does rollover affect future workload and resource availability? Planners need to update schedules dynamically, balancing new and rollover work.
- **Prioritisation:** Some rollover tasks remain critical and must be expedited, while others may be deferred safely.
- **Communication:** Field teams, supervisors, and operations must all be informed of changes to avoid surprises.
- **Root cause analysis:** Frequent rollovers could signal systemic issues - inaccurate labour estimates, unrealistic scheduling, or resource shortages - that planners should investigate.

Maintaining a clear dashboard or report on rollover work can help planners monitor trends and act early to prevent bottlenecks.

The Planner's Role Beyond Scheduling

Daily execution and follow-up highlight the planner's role as a bridge between strategy, field technicians and operations. Staying connected, facilitating communication, and proactively managing changes are what turn good plans into successful outcomes.

By keeping a finger on the pulse of the shop floor, planners not only help maintain equipment reliability but also build stronger relationships with their teams - a vital ingredient for sustained maintenance excellence.

Collaborating with Operations and Technicians: Building Partnerships That Drive Success

Effective scheduling and maintenance execution don't happen in isolation. One of the planner's most critical skills is the ability to build strong, collaborative relationships with both the operations teams and the technicians who carry out the work. Without this buy-in, even the best plans can stall or fail.

Requesting Planned Equipment Downtime: The Art of Negotiation

A cornerstone of preventive and predictive maintenance is having planned equipment downtime. This can be a tricky conversation because downtime means lost production, which operations naturally want to minimize.

Here, the planner's role extends beyond scheduling - it involves influencing and negotiation:

- Early engagement: Engage with operations as early as possible when planning maintenance that requires downtime. The earlier the conversation, the easier it is to find mutually agreeable windows.

- Data-driven justification: Use reliability data, risk assessments, and cost of failure analyses to demonstrate the value of planned downtime versus reactive breakdowns. Numbers and facts go a long way in persuading operational stakeholders.
- Flexible options: Propose alternative windows or split downtime into smaller, less disruptive segments when possible. Showing willingness to accommodate operational constraints builds trust.
- Clear communication: Frame the discussion around shared goals: safety, uptime, cost reduction, and quality. Avoid jargon and tailor messaging to what matters most to the operations audience.

Influencing Operations: Becoming a Trusted Advisor

Planners who build credibility with operations can become trusted advisors rather than just "the maintenance people." This influences not only scheduling but also how operations perceive maintenance overall.

Ways to build this influence include:

- **Delivering results:** Consistently planning work that is completed safely, on time, and with minimal disruption.
- **Transparency:** Being upfront about risks, trade-offs, and consequences of delaying maintenance.
- **Education:** Helping operations understand how proactive maintenance supports production goals, reduces emergency breakdowns, and extends asset life.
- **Involving them in planning:** Inviting operations to planning meetings or shutdown preparations encourages joint ownership of outcomes.

Helping Operations Understand the Value of Proactive Maintenance

Operations teams are often measured by their output and may see maintenance as a necessary evil or interruption. Changing this perception takes ongoing effort:

- **Share success stories:** Highlight examples where planned maintenance prevented costly failures or production losses.
- **Visual aids:** Use dashboards or CMMS reports showing improved uptime, reduced reactive work, and cost savings.
- **Common language:** Translate maintenance terminology into operations' business language - uptime %, cost avoidance, safety incidents avoided.
- **Celebrate wins together:** Recognise collaborative efforts publicly, reinforcing maintenance and operations as partners rather than adversaries.

The Technicians: Key Allies in Execution

Collaboration with the technicians who execute work orders is equally important:

- **Job plan involvement:** Where possible, involve technicians in reviewing or developing job plans. This improves task accuracy and buy-in.
- **Feedback loops:** Encourage and act on technician feedback to continuously improve work instructions and scheduling practices.
- **Respect and trust:** Recognize their frontline expertise and challenges. A planner who listens and responds earns respect and smoother execution.
- **Clear communication:** Provide technicians with timely, detailed information about scheduled work, priorities, and changes.

At its core, maintenance planning is about people as much as processes. Building strong, respectful, and communicative relationships with both operations and technicians unlocks smoother scheduling, better adherence to plans, and ultimately, greater equipment reliability and uptime.

A planner who masters collaboration turns scheduling from a logistical hurdle into a strategic advantage for the entire site.

Key Takeaways

- Yearly PM plans provide structure and ensure all regulatory, safety-critical, and reliability-centred maintenance is scheduled throughout the year.
- Bundle PM tasks strategically to reduce equipment downtime, increase technician efficiency, and consolidate resource usage-without overwhelming field teams.
- The weekly planning cycle is the heartbeat of execution-focus on preparing fully scoped, materially ready, and prioritised jobs for inclusion in each week's schedule.
- Balance required vs available labour by factoring in shift availability, absences, and overtime. Use tools like heatmaps or CMMS schedules to visualise workload.
- Make the plan achievable. Avoid overloading the schedule-under-promising and over-delivering is better than losing credibility with an unrealistic plan.
- Assign jobs smartly in the CMMS-planners assign to shifts; team leaders assign to individuals. This keeps flexibility while ensuring clear ownership.
- Prepare the team in advance. Share schedules at least 48 hours before the work week and hold planning meetings to align the team on priorities and risks.
- Shutdowns require project-style planning. The planner's role spans coordination of tasks, resources, sequencing, materials, communication, and safe restart.
- Use visual tools-Gantt charts, task zones, and command centres help bring clarity during complex, high-risk shutdowns.
- Daily execution support matters. Stay connected to the shop floor, facilitate daily stand-ups, and adjust plans based on real-time constraints.
- Rollovers must be managed carefully. Monitor incomplete work, communicate adjustments, and use rollover trends to spot deeper planning or resourcing issues.

- Collaboration is key. Work closely with operations to secure downtime, justify maintenance windows, and align goals.
- Earn influence through trust and consistency. Planners who deliver results and communicate clearly become valued partners, not just schedulers.
- Field technicians are vital partners. Involve them in planning, respect their input, and respond to their feedback to continuously improve the schedule and job quality.

Chapter 11

Contractor Management

"A contractor is only as good as the scope and support you give them"

Maintenance Planners play a pivotal role in coordinating third-party contractors, yet this responsibility often sits in a grey area - not quite fully owned, yet always critical. Whether it's a specialised vendor for condition monitoring or a full shutdown crew, your ability to successfully manage contractors as part of the maintenance ecosystem can make or break your schedule, budget, and quality goals.

This chapter isn't a comprehensive guide to all contractor management principles, but rather a focused look at how planners can leverage external support effectively - while avoiding common traps.

When to Use Contractors Instead of In-House Teams

Planners are often asked to decide or advise on who should do the work - your internal team or an external contractor. The right decision depends on various factors: urgency, complexity, internal capability, availability, current backlog, and cost.

Situations where contractors are typically used:

- Specialised tasks your in-house team isn't trained or certified for (e.g. vibration analysis, arc flash studies, HVAC gas recharging).
- Peak load situations like shutdowns, when workload exceeds internal capacity.
- Emergency breakdowns where local support is faster than mobilising internal teams.
- Project-based work or installation jobs that are outside standard maintenance scope.
- Geographically distant sites or unmanned facilities.

Situations better suited for in-house execution:

- Routine PMs and corrective work already scoped and scheduled into team workload.
- Tasks with critical knowledge of local systems/processes, where internal history adds value.
- Jobs requiring deep CMMS documentation or feedback, which contractors may not reliably capture.

Be clear about the why behind using a contractor. If it's simply due to a lack of planning or internal availability, that's not sustainable. Consider whether it's a gap in capability, capacity, or training - and document this to support long-term decisions.

The Pros and Cons of Using Contractors

Knowing when and how to use contractors means being aware of the **trade-offs**. Here's a breakdown:

Pros	Cons
Access to specialist skills and equipment	Can be more expensive than in-house
Fast scaling of resource during shutdowns or outages	Inconsistent quality if not well-managed

Helps maintain schedule compliance during high demand	Less invested in long-term system reliability
Keeps in-house team focused on core tasks	Poor documentation if not enforced
No need to carry overhead for rarely-used skills	Can disrupt culture or safety norms if not onboarded well

One of the most underestimated downsides of excessive contractor use is the erosion of technical ownership. When external vendors “own” critical systems over time, internal knowledge degrades, making the site more dependent and vulnerable long-term.

Managing Contractor Costs

Cost control is a key planner responsibility, especially when you’re the one requesting or justifying contractor work. Even though you may not hold the purse strings, your planning and documentation directly influence contractor efficiency and cost containment.

Planner’s role in cost management:

- Define clear scopes of work - avoid vague “inspect and report” tasks unless that’s genuinely the goal.
- Include time estimates and expected labour hours - this keeps contractor invoicing transparent.

- Group similar jobs - batching work during a single site visit reduces mobilisation charges.
- Ensure access to equipment and materials - time spent waiting equals billable idle time.
- Audit invoices - cross-check billed hours with work order timestamps and access logs if available.

Always include start/stop times in work order templates and require that contractors sign in/out of the CMMS, security log, or permit system. This creates a paper trail for invoice validation.

Maintaining Maintenance Quality Standards

Contractors may get the job done, but was it done to your site's standards? That's the planner's job to define and enforce, especially where CMMS records, SOPs, or regulatory compliance are involved.

Quality risks with contractors:

- Minimal CMMS updates (no work notes, no parts usage, no failure codes)
- Skipping standard steps or measurements (especially if not monitored)
- Unsafe practices if unfamiliar with site-specific procedures
- Use of incorrect or unapproved materials or spares

How Planners can help uphold standards:

- Include SOPs and checklists in the job plan shared with the contractor
- Require pre-job briefs or signoffs with the site lead or maintenance supervisor

- Design job plans with mandatory data fields - e.g. meter readings, inspection findings, failure modes
- Set expectations for post-job documentation: who enters it into the CMMS, what's required, and when
- Follow up on completed work - even just a spot check builds accountability

You might not have authority over contractors directly, but your ability to set expectations, document requirements, and involve operations in pre-job briefs gives you practical influence without needing formal control.

Onboarding & Permit Integration

A bonus role planners often hold is making sure contractors are ready to work safely and legally before they arrive.

Planners may be involved in:

- Ensuring contractor pre-qualification and site induction is completed
- Scheduling or coordinating required permits
- Linking permits to CMMS job plans or referencing them in the work order

This isn't always formalised, but in lean teams, it's often the planner who's holding all the moving pieces together - making sure the contractor's tools, paperwork, and access are all in sync with the schedule.

Just like with internal teams, contractor outcomes are directly tied to the clarity and quality of your job plans. The more detailed, scoped, and documented the task, the more likely it is to be delivered right the first time, regardless of who does it.

Think of contractors as a force multiplier - they help you extend your team's capabilities, hit deadlines, and achieve reliability goals. But they only work in your favour when well-managed. Your ability to influence their effectiveness, even without direct authority, is part of what makes a maintenance planner indispensable.

Vetting Contractor Quality: A Planner's Quick Guide

While formal procurement and health & safety teams often handle contractor onboarding and compliance, planners are uniquely positioned to vet and assess contractors from a practical, on-the-ground perspective. You're the one who sees whether the work gets done - properly, efficiently, and reliably.

Here's a streamlined 5-step vetting process that planners can use informally (or formally, if your company supports it) to evaluate contractor quality before work begins and after it's complete.

1. Review Relevant Experience and Past Performance

Before assigning work, ask:

- Have they worked on this type of equipment or system before?
- Do they have references from similar industries or facilities?
- Can they share sample documentation from previous work orders (job reports, photos, data logs)?
- Have they worked at your site before - if so, how did it go?

Use your CMMS history. Check past work orders completed by the contractor for quality of closure notes, timeliness, and whether follow-up work was required.

2. Check Certifications and Technical Competency

While some of this is owned by your HSE or procurement team, as a planner you can:

- Confirm technician-level certifications for the specific work being done (e.g., confined space, electrical licenses, thermography level 1/2/3).
- Ask for copies or confirmation from the vendor directly when planning specialised jobs.
- Ensure the contractor's nominated personnel match the qualifications required in your job plan.

Consider adding a checklist item in your job planning template for "Contractor Certification Verified."

3. Assess Workmanship and Documentation Standards

After an initial job or trial:

- Were their closing notes clear, accurate, and timely?
- Did they record part numbers, tools used, time spent, and findings?
- Was the job left in a clean, safe, and complete state?
- Were any standards, SOPs, or compliance checks skipped or ignored?

Flag poor workmanship immediately and create a review loop with your maintenance manager or contractor liaison. Don't let poor habits embed early.

4. Evaluate Communication and Professionalism

Soft skills matter when contractors interact with your teams and site processes:

- Were they responsive to queries and flexible to scheduling changes?

- Did they attend pre-job briefs or planning meetings punctually?
- Did they raise issues early, or wait until post-failure or follow-up?
- Were they respectful and collaborative with internal technicians?

Good contractors act like an extension of your team. If they stonewall or go silent when something goes wrong, that's a red flag.

5. Capture Feedback and Build a Contractor Scorecard

After each job or work campaign:

- Log feedback from the planner, supervisor, and technicians who worked alongside them.
- Consider building a simple "Contractor Scorecard" template to record:
    - Technical capability (1–5)
    - Safety and compliance (1–5)
    - Communication and responsiveness (1–5)
    - Documentation and CMMS accuracy (1–5)
    - Would you use again? (Yes/No)

Over time, this becomes a powerful internal record of which vendors consistently deliver - and which need corrective action or replacement.

A contractor's reputation should live in your team's planning toolkit - not just in procurement files. This feedback loop is your defence against poor quality creeping into your maintenance program.

Contractors can be valuable partners, but your job as a planner is to standardise their contribution to match your expectations - not lower your expectations to match theirs.

Vetting isn't just about finding the best; it's about setting the bar and giving them a fair chance to meet it. Document what good looks like, enforce it in your job plans and close-out requirements, and use consistent feedback to reinforce long-term reliability.

Key Takeaways

- Planners are central to contractor coordination, even if they don't formally "own" the vendors. Your influence comes through preparation, clarity, and follow-up.
- Know when to use contractors vs. internal teams. Base the decision on complexity, urgency, and capability-not just convenience or short-term gaps.
- Balance the pros and cons of outsourcing. Contractors bring speed and flexibility, but can risk quality, cost, and long-term technical ownership if not managed carefully.
- Effective scope definition and job planning are your best tools for controlling contractor cost, quality, and efficiency.
- Insist on documentation and CMMS accuracy. Make it a non-negotiable expectation for all contractor work-just like you would for internal teams.
- Collaborate with procurement and safety teams, but take ownership for aligning contractor work with maintenance strategy, planning tools, and operational schedules.
- Use informal vetting and feedback loops to build a working "contractor scorecard," tracking who performs reliably and who doesn't meet expectations.
- Standardise expectations. Clear scopes, checklists, and pre-job briefs help ensure contractors deliver to your standards, not just to minimum requirements.
- Treat contractors as temporary extensions of your team, not outsiders. Integrate them thoughtfully, hold them accountable, and protect your maintenance culture and standards.

PART III – Driving Improvement

Chapter 12

KPI's and Metrics That Matter

"Don't let metrics become the goal - let them tell the story"

What Are KPIs and Why Are They Valuable?

Key Performance Indicators (KPIs) are measurable values used to assess how effectively a team, function, or process is performing against its objectives. In the context of maintenance planning, KPIs serve as the dashboard for reliability, efficiency, and effectiveness.

For planners, KPIs provide:

- Clarity on whether the maintenance strategy is working.
- Feedback on how well planning and scheduling processes are functioning.
- Justification for decisions and resourcing.
- Focus for continuous improvement efforts.

From a planner's perspective, well structured KPIs allow you to tell a story: where your operation has been, where it's trending, and where adjustments might be needed. It should not just be about hitting numbers; it is primarily about understanding the performance landscape.

Some common value-adding KPIs in maintenance include:

- Schedule Compliance
- Mean Time Between Failures (MTBF)
- Mean Time To Repair (MTTR)
- Work Order Closure Quality
- Backlog Age or Volume
- Planned vs. Unplanned Work Ratio
- First-Time Fix Rate

We'll cover more of these and how to interpret them throughout this chapter.

The Risk of Overusing Metrics

While KPIs are important, too many metrics can quickly become a distraction or, worse, a form of waste.

It's not uncommon to see dashboards bloated with 20+ metrics, most of which no one looks at or understands clearly. This kind of over-reporting can create:

- Noise: Dilutes the signal of genuinely useful data.
- Confusion: Teams unsure what to prioritize.
- Bureaucracy: Wasted time gathering and reporting for the sake of it.

Ask yourself, "What decisions can I make with this data?" If the answer is "none," it's not a key performance indicator, it's just noise.

Focus on the vital few. Choose 5–7 key metrics that align with your department's goals and that you can genuinely influence through planning, scheduling, and communication. These KPI's can always be amended as time passes and you identify problem areas worth gathering performance data on.

Tracking and Trending Data

A metric is only useful when viewed over time.

Trending is the process of plotting KPIs on a regular cadence - daily, weekly, monthly - to identify patterns. This helps planners:

- Detect slow declines in performance before they become crises.
- Understand seasonal or production-related variability.

- Provide hard data for business cases (e.g. requesting a new hire, changing a PM frequency).

For example: If your Planned vs. Unplanned Work ratio is improving month-on-month, it validates your planning maturity. But if it slips back, it can point to a creeping reactive culture, unexpected asset behaviour, or reduced scheduling discipline.

Some ways to trend effectively:

- Use line graphs or heatmaps for visibility at a glance.
- Automate reporting where possible through your CMMS or BI tools.
- Include a commentary or narrative each month - what changed, and why?

Avoiding "Metric as a Target" Pitfalls

A common failure with KPIs is when the metric becomes the target, rather than being a signal of performance.

This concept is best summed up by Goodhart's Law:

> "When a measure becomes a target, it ceases to be a good measure."

Here are a few ways this often manifests in maintenance:

- Teams closing work orders prematurely just to improve schedule compliance.
- Padding job durations to "always hit estimates."
- Avoiding logging repeat failures to protect MTBF stats.

If your KPI system rewards appearances rather than outcomes, it actively works against your reliability goals.

Metrics should be an output, not an input. They reflect what's happening; they should not dictate short-sighted behaviour. Your role as a planner is to:

- Use metrics as a diagnostic tool, not a scoreboard.
- Look for sustainable improvements, not quick fixes.
- Encourage open dialogue - "what is this data telling us?" rather than "how do we hit this number?"

Useful KPIs for Maintenance Planners

The following KPIs are particularly relevant to the planner's role. They offer insight into the quality of your planning, the efficiency of maintenance execution, and the health of your broader maintenance program. For each one, we'll look at:

- What it measures
- Why it matters
- How to use it
- Common pitfalls to avoid

PM Schedule Compliance

What it is:
The percentage of planned preventive maintenance (PM) tasks that are completed on time, according to schedule.

Why it matters:
High PM compliance reflects strong planning and scheduling practices and helps ensure equipment remains in a reliable, known state. Low compliance usually means you're slipping into a reactive culture - often due to firefighting or overloaded schedules.

How to use it:

- Track monthly or weekly.
- Use missed PMs to trigger review: Was the task still needed? Was it poorly scheduled? Were resources unavailable?
- Break it down by area or system to spot bottlenecks.

Pitfall:
Be cautious of false compliance - e.g. PMs being marked “complete” without actually being executed, just to hit the number.

You own this KPI. PM compliance is the heartbeat of proactive maintenance - and your plan should reflect what’s realistically achievable.

Wrench Time

What it is:
The percentage of a technician’s day spent on actual hands-on maintenance work, as opposed to waiting on parts, walking to job sites, attending meetings, or doing admin.

Why it matters:
Low wrench time is usually not a technician issue - it’s a planning issue. Planners drive wrench time by ensuring:

- Clear job plans
- Pre-staged materials
- Logical job sequencing
- Removal of avoidable delays

How to use it:

- Conduct periodic time studies (shadowing or sampling).
- Use it as a leading indicator for planner effectiveness.
- Investigate causes of low wrench time and correct them systematically.

Pitfall:
Don't weaponise this metric - it's meant to highlight systemic inefficiencies, not micromanage technicians.

Work Order Backlog Health

What it is:
The volume and age of open Work Orders - particularly in the "Ready to Schedule" and "Awaiting Parts" buckets.

Why it matters:
An out-of-control backlog means jobs are falling through the cracks, risks are growing, and planning isn't keeping pace with workload. A healthy backlog:

- Reflects upcoming demand
- Is appropriately prioritised
- Contains only well-scoped, ready-to-execute jobs

How to use it:

- Monitor backlog size vs. available labour.
- Categorise by age and priority (e.g. how many urgent jobs are over 14 days old?)
- Review regularly with maintenance and ops leadership.

Pitfall:
Don't just track the number of WOs - track value and urgency. Not all backlogs are bad; unprioritized or stale ones are.

Spare Part Stock Health

What it is:
A measure of whether your spare parts inventory supports the maintenance strategy effectively, without overstocking or risking critical shortages.

Why it matters:
Poor parts availability leads to delays, rescheduling, and reactive maintenance. Meanwhile, overstocking ties up capital and increases holding costs.

How to use it:

- % of work orders delayed due to unavailable parts.
- Stockout rate for critical spares.
- Value of obsolete/unused inventory.
- Accuracy of min/max levels vs. actual usage.

Pitfall:
This is a shared KPI - planners should work closely with the stores/procurement team to influence it but not try to own it in isolation.

If a job is delayed due to missing parts, ask yourself - was the part unavailable, or was the part not requested when the job was planned?

Equipment Reliability & Availability (Uptime)

What it is:
These are often tracked at a higher level (by Reliability or Ops), but planners play a huge supporting role.

- Availability: % of time the equipment is operational and not down for planned/unplanned maintenance.
- Reliability (MTBF): Average time between failures - signals asset stability.

Why it matters:
Good planning helps avoid unplanned downtime. Poor planning (or lack of it) increases it.

How to use it:

- Look at trends by asset, department, or system.
- Cross-reference unplanned downtime events with missed or poorly executed PMs.
- Use reliability data to justify PM optimisation or condition-based strategies.

Pitfall:
Don't own this metric entirely - but own your contribution. Show how planning is part of the reason uptime is improving.

Expected vs. Actual Labour Hours

What it is:
A comparison between how long a task was planned to take vs. how long it actually took.

Why it matters:
Accurate job duration estimates lead to realistic schedules, balanced workloads, and better resource planning.

How to use it:

- Regularly review "Planned vs. Actual" time on closed work orders.
- Investigate significant deviations.
- Use feedback from technicians to adjust future estimates.

Pitfall:
Avoid blaming technicians when actual time is higher - often it's a result of inaccurate scoping, unclear job plans, or unknown complications.

Other Useful KPIs

While the ones above are some of the most important, planners may also benefit from tracking:

KPI	Description
First Time Fix Rate	% of jobs completed without needing rework or follow-up
Job Plan Reuse Rate	% of recurring work using standard plans
Work Order Closure Time	Time between completion and formal closure (affects data quality)
% Work with Complete Data	How many WOs include accurate time, parts used, failure code etc.
% Planned Work	What % of total maintenance work was planned (vs. reactive)

KPIs Are a Mirror, Not a Hammer

KPIs should help you see clearly - not beat people over the head.

> “What gets measured gets managed… but only if it’s understood, relevant, and acted upon.”

As a planner, you play a central role in shaping the maintenance team’s data. You:

- Influence the input (via job planning)
- Shape the process (via scheduling and coordination)
- And enable insights (by ensuring work orders are closed properly and meaningfully)

Used well, KPIs can validate your hard work, reveal opportunities, and demonstrate the value that effective planning brings to a reliability-centred organisation.

Measuring, Tracking & Influencing KPIs: From Data to Decisions

The Purpose of Tracking: Seeing the System

Before getting into dashboard tools or trend lines, it's worth revisiting why we track KPIs in the first place. KPIs aren't about creating extra admin or satisfying a leadership curiosity - they're there to shine a spotlight on the health of the system you're working in.

A good KPI gives you visibility into:

- Process effectiveness (e.g. how well job plans are executed)
- System balance (e.g. is your backlog growing faster than you can manage?)
- Team performance (e.g. wrench time and schedule compliance)
- Business alignment (e.g. is your planning supporting uptime goals?)

In other words: metrics are a feedback loop, and the planner is both a consumer and creator of the data that drives that loop.

Designing and Building Performance Dashboards

Whether you're using CMMS-native reporting tools, Excel, Power BI, Tableau, or another business intelligence platform, the same principles apply when building performance dashboards that are useful, not just pretty.

Key Design Principles

1. Audience clarity
    - Build different views for different roles.
      → Technician-level dashboards should be simple, task-focused.
      → Leadership dashboards should align with strategic outcomes (uptime, cost, risk).
2. Keep it tight
    - Focus on 5–8 key metrics per view. Any more, and you're diluting insight.
3. Visual cues matter
    - Use colour, traffic lights, trends over time, and simple visuals to quickly communicate where attention is needed.
4. Enable drill-downs
    - A good dashboard lets you zoom in on an issue - e.g. seeing which PMs were missed, or which assets are causing the most downtime.
5. Time-frame awareness
    - Make sure it's clear whether you're looking at:
        - Real-time data
        - Rolling weekly/monthly averages

Dashboards don't need to look fancy - but they must be actionable. Ask yourself: "Does this help me make a better decision this week?"

Using the CMMS as a Source of Truth

This is where earlier chapters connect back: your CMMS is only as useful as the quality of data put into it.

- If techs don't close jobs properly, labour hours and parts consumption are inaccurate.
- If failure codes aren't used, you can't trend reliability.
- If follow-up WOs aren't created, recurring issues are missed or misattributed.
- If job plans aren't reused or updated, planning accuracy becomes stagnant.

As a planner, your role is to champion the value of good data. That doesn't mean becoming a data police officer - it means showing your team how quality inputs improve job prep, make their lives easier, and reduce fire-fighting.

Practical Tips:

- Lead by example: Ensure your own work orders are complete and clean.
- Create quick "WO closure guides" for field teams if needed.
- Review a few WOs each week in team meetings - call out what's done well.
- Provide feedback loops. If a tech's detail helped prevent rework or justified a part reorder, tell them.

Trending Metrics Over Time

KPIs are much more powerful when trended, not just snapshot-reported.

Examples:

- PM Compliance trend: Are we improving, slipping, or stable over time?

- Planned vs. actual labour hours trend: Are our estimates becoming more accurate?
- Backlog age buckets trend: Is overdue work growing or shrinking?
- Spare part stockouts trend: Are we getting better at forecasting?

Why it matters:

- Spot systemic issues early
- Reveal the impact of process improvements (or their absence)
- Justify resource requests (e.g. "We've seen a 20% increase in planned hours/month - we need to reallocate tech time.")

Don't just report data - tell a story. Use trends to explain what's changing, why it matters, and what's being done about it.

Influencing Performance with Metrics

Numbers alone don't improve performance - your actions and communication do. Use your dashboard to create a culture of continuous improvement.

Ways to Influence:

- Share metrics openly during stand-ups or weekly planning meetings.
- Translate KPIs into technician impact:
    - "Because we hit 95% PM compliance last month, we had 40% less emergency work."
    - "Your detailed comments helped us solve a recurring motor issue faster."

- Celebrate wins - especially when data shows a direct link to technician effort or improved planning.
- Use metrics in coaching, not punishment.
- Tie planning KPIs into bigger conversations with Ops and Reliability - show how your work supports site goals.

Linking Planning KPIs to Business-Level Outcomes

The best planners don't just think in terms of wrench time or backlog - they think in terms of business impact.

Here's how common planning metrics tie into broader goals:

Planning KPI	Business Outcome Supported
PM Compliance	Equipment reliability, uptime, reduced product loss
Planned vs Actual Labour	Improved cost control, reduced overtime
Work Order Closure Quality	Better failure analysis, reduced repeat breakdowns
Wrench Time	Increased asset utilization, less production disruption

Spare Part Stock Health	Inventory cost control, reduced working capital
Backlog Health	Lower risk of unexpected failures or regulatory gaps

Planners who understand and speak the language of business outcomes gain greater influence in their organisations. You're not "just" managing jobs and parts - you're helping protect uptime, reduce cost, and ensure product quality.

If your team is facing pressure to reduce costs or improve reliability, look at your dashboard. Which lever can you pull this month that helps move the needle?

Key Takeaways

- Data is only valuable when it drives action. Maintenance KPIs should be selected and reported in a way that leads to better decision-making, not just dashboard filling.
- The planner plays a vital role in data quality. Poor job close-outs, vague failure codes, and missing time entries compromise the integrity of your metrics - planners must lead in setting and enforcing data standards.
- Core maintenance metrics include schedule compliance, PM completion rate, backlog size and age, wrench time, and mean time between failure (MTBF) - but context matters more than raw numbers.
- Visualise data clearly and consistently. Use tools like Power BI, Excel, or built-in CMMS dashboards to create views that highlight trends, exceptions, and opportunities for improvement.
- Track leading and lagging indicators. Leading indicators (like planner workload, job readiness, and schedule loading) help you prevent issues. Lagging ones (like downtime and failure rates) show results after the fact.
- Don't track KPIs in isolation. Cross-reference metrics (e.g. PM compliance vs. reactive work rate) to spot underlying issues and validate the effectiveness of your planning strategy.
- Share reports openly with your team and stakeholders. Make your maintenance metrics part of the conversation - not hidden in spreadsheets or leadership decks.
- Planners should become data storytellers. Learn to interpret what the numbers mean, explain them clearly, and turn them into compelling reasons for change or improvement.

Chapter 13

Continuous Improvement

"Improvement doesn't need permission - it needs observation, action, and follow-through"

Why Continuous Improvement Matters

Continuous Improvement isn't a slogan. It's a mindset, a discipline, and a core function of world-class maintenance planning. In fact, if you're doing your job well as a planner, you're already in the business of improving - even if you don't realise it yet.

At its core, continuous improvement is about eliminating waste, reducing variation, increasing reliability, and ultimately driving more value from the maintenance effort. That value can take many forms: fewer unplanned breakdowns, better technician productivity, lower spare part costs, or more efficient planning processes. Over time, these gains compound, and they're usually driven not by one big change, but by hundreds of small, deliberate improvements.

So why is this particularly relevant to the planning function? Because the planner is often the person closest to the systems, data, and processes that can either support or hinder improvement. You see the gaps in job plans. You should notice repeated breakdowns. You hear the feedback from technicians. You spot the spare part ordering mistakes. You sit at the intersection of planning, execution, and performance. That's exactly where improvement starts.

The PDCA Cycle: A Framework That Fits

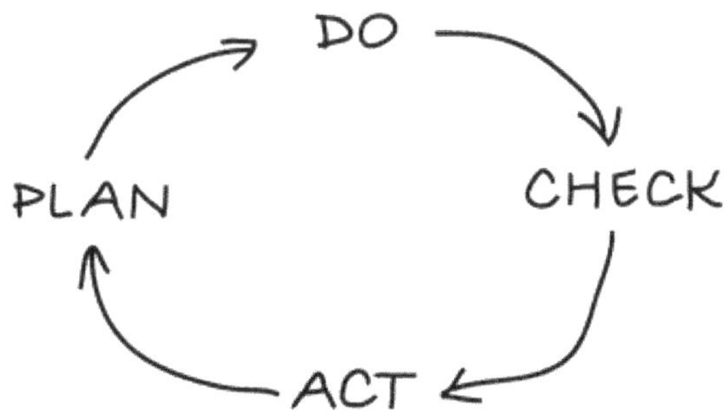

The Plan-Do-Check-Act (PDCA) cycle - sometimes called the Deming Cycle - is a widely used framework for driving structured, repeatable improvement. It's deceptively simple, but incredibly powerful when applied in the maintenance planning world.

Let's break it down through the planner's lens:

PLAN

This is where you define the problem or opportunity, analyse data, and develop a hypothesis for improvement.

As a planner, this could mean:

- Noticing that a particular asset has failed three times in six months.
- Reviewing the work order history and spotting that the job plan is vague or missing key checks.
- Working with a technician to develop an improved task sequence or inspection step.

- Or it could be something broader - like identifying that backlog is building due to under-resourcing during holiday periods, and proposing a scheduling adjustment.

Planning isn't just about managing work - it's about identifying patterns, proposing changes, and setting the conditions for success.

DO

This is the implementation step - not of the full-scale change, but of a controlled test or first iteration.

For example:

- You trial the improved job plan on the next two planned interventions.
- You implement a new template for capturing WO failure data on a specific asset type.
- You adjust the weekly plan to reserve more capacity in December and monitor impact on backlog.

The key here is to execute intentionally and gather data, not just make ad-hoc changes.

CHECK

This step is often overlooked - but it's where the learning happens. What does the data say? What was the result of the change?

Some ways you might check:

- Compare job completion times before and after a job plan improvement.
- Review technician feedback on the new task format.
- Look at MTBF (Mean Time Between Failures) on an asset before/after changes.

- Use your CMMS to review backlog trends, labour utilization, or PM compliance.

Improvement without measurement is just guessing. A great planner is also a feedback-looper.

ACT

Now it's time to standardise the successful change - or, if it didn't work, tweak and retry.

That might mean:

- Rolling out the updated job plan site-wide.
- Updating your team's closure procedure for WOs.
- Making a business case for a permanent scheduling change.
- Or, if your trial didn't yield results, revising your hypothesis and returning to the Plan stage.

The beauty of the PDCA cycle is that it never really ends - it's a habit, not a one-time initiative. And in a mature planning environment, you'll be cycling through dozens of these mini PDCA loops every year - some formal, some informal, all pushing performance in the right direction.

Root Cause Analysis (RCA)

What Is Root Cause Analysis?

Root Cause Analysis (RCA) is a structured process for identifying the fundamental reason why a problem occurred. In the maintenance world, that usually means looking beyond the obvious symptoms of a failure - such as "a motor stopped working" - and asking, *why did this happen, and what can we do to stop it happening again?*

From a planner's point of view, RCA is an essential tool not just for solving problems, but for learning from them. It's about turning reactive work into proactive knowledge.

Why Is RCA Valuable for Maintenance Planners?

Planners are in a uniquely powerful position when it comes to RCA. You have visibility of:

- Work order histories
- Equipment performance trends
- Technician feedback
- Job plan and task list content
- Parts usage and replacements

This broad view means you can spot patterns others might miss, and initiate or support RCA efforts before the same failure repeats.

Also, a good RCA often triggers improvements to:

- PM content and frequency
- Job plans and task steps
- Spare part specifications
- Technician instructions or SOPs

All of which fall within the planner's domain, so you can see the value to be gained from carrying out full and proper root cause analysis following unexpected problems and failures.

The 5 Whys Technique

One of the most accessible and commonly used RCA tools is the **5 Whys** method. It's exactly what it sounds like: a structured series of "why" questions, where each answer leads you to a deeper cause.

Example:

- **Problem:** A gearbox failed during production.
- **Why?** → The gearbox overheated.
- **Why?** → The oil level was too low.
- **Why?** → The oil seal had worn and was leaking.
- **Why?** → The seal was not replaced at the last PM.
- **Why?** → The PM checklist didn't include an inspection of the seal.

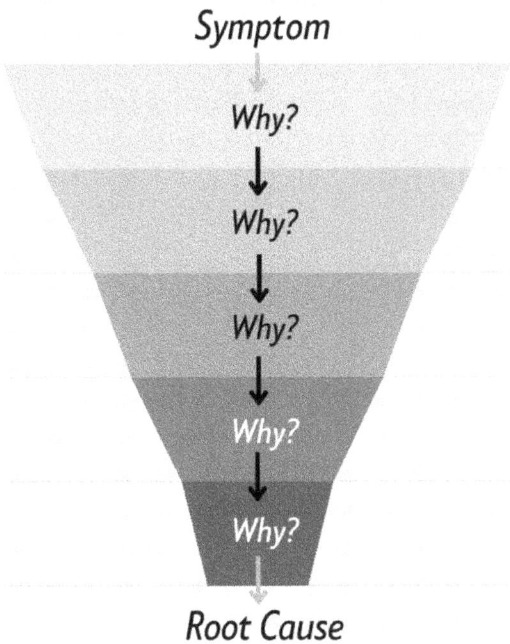

Root Cause of the Gearbox failure: Incomplete PM checklist - a planner-controlled document.

This kind of insight allows you, the planner, to directly improve the system by updating the PM instruction. A £10 seal added to a task plan might prevent a £10,000 failure.

What could have happened in this example without the 5 Why analysis, is the gearbox failure could be dismissed as just end of life or random failure, without taking the learnings which could prevent it happening again.

When Should RCA Be Used?

This is an important question. **Not every failure deserves a full-scale RCA.**

Sometimes, the failure mode is obvious and non-repeating. Sometimes, the cost of analysis outweighs the impact of the issue.

Here's a simple rule of thumb to help decide when RCA is appropriate:

Criteria	RCA Recommended?
Safety incident or near miss	✓ Always
Major equipment failure (>£5k or critical asset)	✓ Yes
Repeat failures of the same type	✓ Yes

Single failure with clear, known cause	✗ No need
Minor one-off issues (e.g. loose bolt)	✗ Possibly not
Customer impact or unplanned downtime	✓ Worth considering

In your role as a planner, you don't always lead the RCA - but you should always be ready to contribute with the information and date you have at your disposal:

- Pulling together work order history
- Providing failure data and technician notes
- Checking what inspections or PMs were in place
- Recommending improvements to the task content

Don't Just Identify - Act

RCA is only as good as the actions it leads to.

As a planner, your job is not just to help uncover the root cause, but to ensure the solution is embedded into the plan.

That might mean:

- Creating a new PM or CBM trigger
- Updating an inspection route
- Adding a torque check to a job plan
- Increasing the frequency of a lubrication task

- Flagging a potential design flaw to engineering

Every RCA should end with a closed loop of learning, documented improvement, and follow-up. And the CMMS should reflect this change - so that future planners, techs, and auditors can see *why* the job plan evolved.

Kaizen in Maintenance Planning: Small Changes, Big Impact

Kaizen - the Japanese term meaning *"change for better"* - is a foundational concept in continuous improvement, especially within lean manufacturing and maintenance environments. While Kaizen often conjures images of cross-functional workshops and structured improvement events, in the planner's world, it can be far more organic.

In the maintenance planning context, Kaizen is about relentlessly tuning your systems - job plans, schedules, equipment metadata, PMs, parts libraries - to better serve the team, reduce waste, and increase reliability.

What Kaizen Looks Like for a Planner

Kaizen in planning is rarely about huge overhauls. It's about paying attention to the friction points:

- Why do techs keep skipping this job step?
- Why do we always run out of this consumable?
- Why does this PM take twice as long as we estimated?
- Why do these breakdowns keep recurring?

Every small adjustment counts:

- Add a missing step to a job plan
- Flag an unstocked part to purchasing

- Update a PM frequency based on new CBM data
- Improve clarity in an SOP
- Reorganise how jobs are grouped on the schedule for less travel time

This kind of incremental optimisation is at the heart of a planner's value-add. You are the team's systems steward - and Kaizen is your core methodology.

Data and Feedback Loops: Turning Metrics Into Improvement

Collecting data is not enough. The true value comes when we feed it back into the planning cycle and act on it. This is the essence of *closing the loop* - and it's where many organisations fail.

Let's break this into a few key principles:

1. Metrics Are Signals, Not Endpoints

As explored in Chapter 12, KPIs like PM compliance, failure rates, backlog size, or wrench time are all useful. But they are *not the objective* - they are the signals that something might need improvement.

For example:

- Low PM compliance → Why? Too many tasks? Wrong time slot? Resource clash?
- High emergency work rate → Are we missing failure modes? Are the right PMs in place?
- Large backlog → Are jobs overestimated? Understaffed? Poor prioritisation?

The planner's role is to interpret these signals and initiate adjustments - to job plans, schedules, staffing models, PM strategies, or parts stocking.

2. RCA → Actionable Change in the System

When we complete an RCA and learn something new, that knowledge must not sit in a report - it must change how we plan.

That could mean:

- Adding a new inspection to a task
- Creating a follow-up CBM check
- Updating equipment metadata to include a newly discovered failure point
- Revising the skill requirements on certain jobs

This is the Kaizen loop in action:

Data → Insight → Action → System Change → New Data

And the planner often acts as the owner of this loop.

3. Use the CMMS as a Feedback Engine

Modern CMMS systems are powerful - but only if used with discipline. Here's how to turn your CMMS into a true continuous improvement platform:

- **Failure codes**: Ensure they are being used consistently. Analyse top 10 causes regularly.
- **Closing comments**: Train technicians to enter meaningful feedback - what worked, what didn't, suggestions for improvement.
- **Actual time vs estimated**: Track discrepancies and refine your planning estimates.
- **Task effectiveness**: Spot which PMs are catching issues - and which aren't.

All of this depends on data quality - which you've covered in earlier chapters. But this is where that investment pays off. *Bad data → no insight → no improvement.*

Establish a monthly review cycle, even an informal one, to look at the top recurring failures, common technician feedback themes, or jobs with time estimate mismatches. Build improvements into the plan every month.

Closing the Loop:

Making Continuous Improvement... Continuous

Here's a visual way to think about the planner's continuous improvement role:

Plan → Execute → Review → Adjust → Plan again

- **Plan**: Build the job, write the PM, assign the schedule
- **Execute**: Techs perform the work and input data
- **Review**: CMMS shows results, failures, time spent, technician feedback
- **Adjust**: You improve the plan, fix job steps, correct spares, modify schedules

Every pass through the loop should tighten your system.

Even better - loop in your field teams. When they see that their feedback leads to better plans, they become more engaged, and the quality of information improves again. That's the virtuous cycle.

PM Quality Auditing: Strengthening the Backbone of Maintenance

Preventive Maintenance (PM) programs are the backbone of many maintenance strategies, but just having them in place isn't enough.

Over time, task drift, equipment changes, evolving failure modes, or poor execution can erode PM effectiveness. That's where PM auditing steps in - a crucial component of continuous improvement, and one that planners are well-positioned to lead.

Why Audit PMs and Job Plans?

PMs are not static - equipment evolves, failure modes change, and the workforce turns over. Without structured review and feedback loops, PM tasks often become:

- Outdated (covering legacy risks or removed components)
- Ineffective (failing to prevent repeat failures)
- Wasteful (duplicating efforts or applying excessive frequency)

Auditing PMs helps ensure:

- Tasks are still relevant and focused on critical failure modes
- Instructions are clear, complete, and executable
- The effort is proportional to the risk and value

This improves both equipment reliability and technician efficiency.

The Planner's Role in PM Auditing

Planners are ideally placed to facilitate and drive the PM audit process - coordinating across engineering, field technicians, and reliability teams.

Key responsibilities include:

- Scheduling periodic reviews of PM tasks (e.g., annual cycle or post-failure analysis)
- Reviewing the structure and content of PMs in the CMMS
- Gathering and incorporating feedback from technicians
- Working with reliability engineers or SMEs to align tasks with known failure modes

- Updating and controlling version history of task plans

Planners act as custodians of task clarity, effectiveness, and efficiency - using audits to tighten the feedback loop between plan and execution.

A Basic PM Audit Checklist for Planners

Here's a simple checklist to use when auditing PM tasks. This can be used as part of a structured review or even as a lightweight review process when techs report issues with a job plan:

Task Content and Clarity

- ☑ Are the instructions clear, step-by-step, and actionable?
- ☑ Are technical terms and abbreviations understandable to all relevant technicians?
- ☑ Is the task written in the correct sequence of execution?
- ☑ Are required readings, measurements, or pass/fail criteria clearly stated?

Tools and Parts

- ☑ Are all required tools listed?
- ☑ Are spare parts or consumables identified with part numbers and stock locations?
- ☑ Are any critical tools (e.g. calibration devices) noted and accessible?

Safety and Compliance

- ☑ Are all relevant safety steps included (e.g., lockout/tagout, PPE)?

- ☑ Are regulatory checks (e.g., GMP, safety inspections) incorporated if required?
- ☑ Are permits or documentation requirements referenced?

Frequency and Duration

- ☑ Is the PM frequency still valid based on equipment performance or history?
- ☑ Does the estimated duration reflect actual time needed to complete?
- ☑ Are there opportunities to combine tasks to improve efficiency?

Field Execution Feedback

- ☑ Has technician feedback been sought and recorded?
- ☑ Have any comments or complaints been addressed in this revision?
- ☑ Are there recurring remarks on specific instructions or parts issues?

This checklist can be built into a formal audit document, or used as a quick evaluation tool when revising existing PMs. Over time, this review process builds maturity into the maintenance program.

Auditing PM Execution

While auditing PM content is essential for improving job plans and instructions, it's equally important to audit the actual execution of PMs on the shop floor. This is where the planner can uncover whether what was planned is translating into quality work - and where subtle gaps in execution standards can quietly erode equipment reliability.

What Field PM Audits Involve:

Field PM audits are quick, structured walk-throughs, typically carried out shortly after a piece of equipment has undergone a scheduled PM. These reviews are often conducted by the planner in collaboration with a shift leader, maintenance supervisor, or reliability engineer. The aim is to assess whether the work was:

- Completed in full, to the defined standard
- Carried out safely and cleanly
- Documented accurately
- Delivering the intended improvement or risk mitigation

Things to Look For:

- Visual inspection of the equipment: Is it clean? Are covers and fasteners secure? Has lubricant been applied or drained where expected?
- Evidence of completed steps: For example, if the PM included belt tension checks or filter replacements, is there evidence it was done?
- Signs of shortcuts or neglect: Missing grease caps, inspection points that clearly weren't accessed, or unchanged filters are red flags.
- Data match-up: Do the completion notes or system entries reflect what you see? Were measurements entered? Do they make sense?

Even with well-crafted job plans, execution can slip. Field audits reveal where:

- PMs are being "ticked off" without real value being delivered.
- Technicians are bypassing key steps due to time pressure or unclear instructions.
- Job plans may be overly generic, impractical, or no longer fit-for-purpose.

- Training gaps exist in understanding the intent or method of the task.

How to Embed PM Execution Audits:

- Schedule monthly or bi-weekly audits, choosing different assets and technicians each time to get a balanced picture.
- Use a simple checklist to record observations (you can build this into your planner toolkit).
- Share findings openly - good and bad. Celebrate clean, complete work. Address recurring gaps in a no-blame way.
- Use feedback to adjust job plans, provide technician coaching, or improve scheduling practices.
- Trend audit data in some way to monitor overall direction of travel - is PM execution generally getting better of worse?

Remember, PM audits are not there to catch people out - they are there to close the loop between planning and execution. Field audits show technicians that what they do matters, and help ensure that PMs are delivering real-world reliability improvements.

Gathering and Using Feedback from Technicians

A successful PM audit process depends on field input. The technicians are the ones executing the tasks, seeing the equipment daily, and knowing whether the plan matches reality.

Ways to gather this feedback include:

- Post-task surveys or digital feedback fields in CMMS
- Toolbox talk sessions focused on task clarity
- Informal check-ins during daily stand-ups
- Including senior techs in formal PM reviews

Most importantly - act on the feedback. When technicians see their comments resulting in real changes, engagement improves and the feedback loop strengthens.

The Value of PM Auditing

Done well, PM auditing delivers:

- Higher task effectiveness - more failures prevented, fewer wasted hours
- Better technician compliance - clear tasks get done properly
- Improved system health - fewer surprises, better uptime
- Reduced risk - from both compliance and operational failure perspectives
- Increased technician trust in the planning process
- Validation that maintenance is being executed to the expected standards

And for the planner, this is one of the best ways to demonstrate strategic value. You're not just pushing jobs into the schedule - you're refining the very core of the maintenance strategy.

Key Takeaways

- Continuous improvement is a mindset, not a one-time project. Planners are uniquely positioned to drive this due to their central role in the data, systems, and feedback loops that underpin maintenance execution.
- Use the PDCA (Plan-Do-Check-Act) cycle to structure and test improvements, from refining job plans to adjusting schedules or PM strategies.
- Root Cause Analysis (RCA) is an essential tool for learning from failures. Techniques like the 5 Whys help uncover system-level issues that planners can fix through updates to PMs, job plans, or spares strategies.
- Kaizen in planning means continuously improving job clarity, workflow efficiency, and data integrity-often through small, consistent actions that eliminate friction or failure points.
- Planners are data translators. They should monitor KPIs, feedback, and failure trends, using this information to drive actionable change in the planning process.
- CMMS feedback loops are critical. Leverage features like failure codes, closing comments, and estimated vs actual labour times to refine your plans and task durations.
- PM auditing is essential. Don't assume a PM is effective just because it exists - routinely review task clarity, structure, frequency, and field execution to ensure your plans are adding real value.
- Audit PM execution in the field, not just on paper. Partner with supervisors to visit assets post-PM, check condition, confirm work quality, and close gaps between plan and reality.
- Technician feedback is a goldmine. Create structured channels to gather it, act on it, and show that their input improves the system-this reinforces engagement and improves long-term outcomes.

Chapter 14

Technology Trends and the Future of Planning

"You don't need to be a tech expert - but you do need to speak the language"

The maintenance planning function is undergoing rapid transformation. As digital technologies evolve and proliferate across operations, planners are expected to do more than just build work orders and chase spare parts. They must now operate within highly integrated systems, interpret increasingly complex data, and help drive digital transformation efforts from the ground up.

But while the tools may change, the planner's core value - translating strategy into structured, executable work - remains as vital as ever. In this chapter, we explore the key technology trends reshaping maintenance planning and the mindset needed to thrive in the years ahead.

The Evolving Tech Stack

Gone are the days when a standalone CMMS and a few paper job cards were sufficient. Today's maintenance technology landscape is increasingly built on interconnected systems - each adding a layer of capability and complexity.

Modern planners are now working within broader Enterprise Asset Management (EAM) frameworks that link maintenance activity to enterprise-wide decision-making. These systems often sit alongside Enterprise Resource Planning (ERP) platforms, connecting finance, procurement, and HR workflows with the maintenance schedule.

Meanwhile, IoT (Internet of Things) sensors and SCADA systems stream real-time equipment data into these platforms, enabling predictive insights and tighter feedback loops. In many organisations, planners are expected to consume, interpret, and act on this information - sometimes in real time.

Just as importantly, the systems themselves are becoming more accessible. Cloud-based CMMS platforms are now the norm, offering mobile-first interfaces, cross-device access, and instant updates. The planner's desk is no longer a fixed location - it's wherever the information and execution needs to happen.

Predictive Maintenance Technologies

Predictive Maintenance (PdM) is one of the most tangible examples of how technology is reshaping the planner's world. While planners may not be vibration experts or thermography analysts themselves, they are increasingly required to understand these techniques and how to integrate them into work planning.

Common PdM tools include:

- Vibration analysis to detect misalignment or bearing degradation
- Infrared thermography to identify electrical hotspots or insulation failure
- Ultrasound to uncover steam trap issues or early-stage bearing failure
- Oil analysis to assess lubricant condition and contamination

When PdM data indicates an anomaly, planners and reliability engineers must act quickly and confidently. That means understanding what the data implies, how urgent the issue is, and how to scope the right corrective job. Going back to the chapter 2 where we explored the skills of a strong maintenance planner - this doesn't necessarily mean you need to be the most technically advanced person (you have reliability engineers in the team for that).

However, you should build yourself a well-rounded understanding of the system, what warning signs to look out for, and what action should be taken in common scenarios.

In mature environments, PdM inspections are structured as routine tasks within the CMMS - and planners are a key component in the scheduling of them, capturing the data, and creating follow-up work orders when thresholds are exceeded.

The planner's role is to be the bridge between the data and the shop floor - translating sensor signals into safe, effective maintenance actions.

Data Skills for Modern Planners

With systems generating more data than ever, one of the most valuable skills a planner can develop is data fluency.

This doesn't mean becoming a data scientist. It means knowing how to extract and interrogate large and complex data sets, interpret trends, spot anomalies, and use the CMMS as a source of insight - not just administration.

For example, planners should be able to:

- Pull reports on overdue PMs or high-reactivity areas
- Analyse failure trends by equipment type or location
- Compare planned vs. actual durations to refine estimates
- Identify where certain job plans lead to frequent follow-up work

When you can connect the dots - not just execute tasks - you become a strategic advisor. You help engineering teams spot reliability risks, operations teams schedule smarter, and leadership teams justify investment in improvements.

Data-informed planners don't just react - they anticipate.

The Shift to Automation and AI

One of the most talked-about trends in modern maintenance is automation. Some CMMS platforms now offer auto-scheduling tools, where the system builds the weekly plan based on resource availability, job readiness, and business priorities. Others incorporate AI-driven failure prediction, surfacing risks based on asset behaviour, historical patterns, or even weather forecasts.

In advanced applications, planners may even interface with digital twins - virtual models of equipment or systems - to simulate work before it's done.

But here's the key truth: automation will not replace the planner.

It might handle some task sequencing, or optimise wrench time, but it won't replace your judgment. Automation can't read between the lines of a vague work request, or negotiate a shutdown with operations, or sense that a technician isn't confident in the plan.

As technology handles the transactional tasks, planners become more valuable - not less. Your role evolves from task manager to reliability enabler, problem solver, and communication hub.

Cybersecurity and Data Ownership

As planners become more digitally connected, they also become stakeholders in data integrity and cyber hygiene. This isn't about technical firewalls - it's about understanding your responsibilities as a data handler.

Maintenance data may include:

- Equipment vulnerabilities
- Contractor and technician identities
- Safety-critical task documentation
- Operational work schedules

All of this can be sensitive. As a planner, ensure you:

- Use secure channels when sharing data
- Respect data governance rules - don't override systems or backdoor access
- Understand who owns the data you enter or extract
- Know what to do if a system seems compromised

Equally, you should be a voice in conversations about data ownership - especially when third-party vendors or cloud platforms are involved. Ask questions: Who can access your job plans? Where is your failure data stored? How long is it retained?

You don't need to be the cybersecurity expert - but you must be part of the conversation.

Preparing for the Future

The technology landscape will keep changing. What's state-of-the-art today will be baseline tomorrow. The best planners prepare not by mastering every tool, but by staying curious, collaborative, and adaptable.

Continuous upskilling is the name of the game. That might mean:

- Completing training on your CMMS's newest features
- Joining webinars on AI in maintenance
- Shadowing your site's PdM specialists to learn the basics
- Asking IT to show you how APIs work or how data flows between systems

The second key is collaboration. Planners who thrive in the future will work closely with IT, operations, engineering, and procurement to ensure the planning process is integrated across disciplines. No more silos; you'll be the connector.

Finally, great planners become change champions. When your site introduces a new system or technology, be the person who helps others adopt it. Translate the jargon. Turn the strategy into action. Build the workflows. Test it, break it, fix it.

That's how planning stays at the heart of reliability - even in a world of rapid change.

Key Takeaways

- CMMS is no longer standalone - Modern planners operate within an evolving digital ecosystem that includes EAM, ERP, SCADA, and IoT platforms. Integration is the new baseline.
- Predictive maintenance is becoming mainstream - Techniques like vibration analysis, ultrasound, and oil sampling are shifting the focus from scheduled to condition-based tasks.
- Data fluency is a core planner skill - It's no longer just about entering data. Planners need to interpret it, spot trends, and use insights to drive smarter planning decisions.
- Automation is accelerating, but human judgment matters more than ever - AI, auto-scheduling, and digital twins will assist, but they can't replace on-the-ground context or cross-team coordination.
- Cybersecurity is everyone's business - As planners interact with increasingly connected systems, understanding data integrity, access control, and compliance is part of the role.
- The planner of the future is a cross-functional collaborator - Working with IT, engineering, operations, and reliability teams is becoming a day-to-day expectation.
- Upskilling is not optional - Tools will evolve, systems will change. The best planners invest in ongoing development to stay ahead and remain relevant.

Chapter 15

Sustainability and Planning for ESG Goals

"Smart planning supports sustainability - one efficient task at a time"

The maintenance planner's role is evolving beyond efficiency, uptime, and cost control; planners today are increasingly expected to contribute to broader organisational objectives - and that includes sustainability. Environmental, Social, and Governance (ESG) goals are now a key part of many companies' strategies, especially in asset-intensive industries. While ESG might feel like the domain of boardrooms and policy teams, planners are uniquely positioned to drive real, on-the-ground progress.

In this chapter, we'll explore how maintenance planners can embed ESG principles into everyday planning work, not as an add-on, but as a natural extension of good practice.

The Planner's Role in ESG Initiatives

While planners likely won't be setting ESG targets, they are often involved in implementing some of the processes that make those targets achievable. A planner who understands how their decisions affect energy use, waste generation, workforce well-being, and regulatory compliance becomes more than just an operations coordinator - they become a value enabler.

This alignment happens through hundreds of small decisions: choosing when to schedule maintenance, which parts to use, how tasks are sequenced, and how assets are recorded and tracked. Every one of those choices has an impact.

Environmental: Planning for a Greener Operation

Environmental sustainability is often the most visible ESG pillar. From reducing carbon emissions to managing energy consumption and waste, the planner has multiple levers to influence environmental outcomes.

Energy-Efficient Scheduling

Poorly maintained equipment consumes more energy - and costs more to run. By maintaining a high standard of preventive maintenance, planners reduce energy waste caused by overheating motors, misaligned belts, blocked filters, or friction losses. More strategically, planners can:

- Schedule high-load equipment PMs during low-demand energy windows
- Cluster maintenance tasks to minimise repeated warm-up cycles
- Work with operations to align with site-wide energy reduction programs

Waste and Spare Parts Management

Inventory control is also an environmental issue. Excess stock risks expiry, degradation, or disposal. Planners can reduce waste by:

- Reviewing parts usage trends and adjusting reorder points to suit demand
- Prioritising repair over replace where feasible
- Coordinating planned overhauls to extend asset life, reducing landfill
- Minimising over-ordering for shut-downs through tighter planning

Supporting Sustainable Procurement

More companies are now specifying that suppliers meet environmental standards. Planners can contribute by flagging sustainable options when specifying parts or selecting vendors.

Where possible, incorporate sustainable criteria into your planning templates or approval workflows.

Social: Planning with People in Mind

The 'S' in ESG often gets overlooked in technical roles, but maintenance work is, at its core, a human process. Planners shape the daily experience of technicians, and the way work is planned affects not just productivity, but morale, safety, and well-being.

Safer Work Through Smarter Planning

Every job plan is an opportunity to reinforce safety. Clear lockout/tagout steps, PPE requirements, or access precautions reduce injury risk. Planners can go further by:

- Designing plans that avoid unnecessary exposure to hazardous conditions
- Rotating repetitive task assignments to reduce strain
- Incorporating post-task safety feedback into job plan reviews

Supporting Work-Life Balance

A planner who balances workloads realistically - avoiding late-notice weekend work or excessive overtime - contributes to employee retention and job satisfaction. Planning with discipline, not just urgency, sends the message that well-being is a factor in operational decisions.

Inclusion Through Documentation

Clear, standardised job plans also reduce barriers for newer technicians or those unfamiliar with specific systems. Good planning creates equitable access to task information, reducing reliance on tribal knowledge.

Governance: Visibility, Accountability, and Compliance

Governance in ESG means ensuring that organisations operate with transparency, consistency, and ethical integrity. Maintenance planning is a key part of that governance infrastructure.

Data Discipline and Traceability

Good planners are data custodians. The CMMS holds audit trails that matter - not just for internal tracking, but for regulatory and third-party reviews. By ensuring job plans, failure codes, and completion notes are accurate, planners provide a governance-ready data stream.

Examples of governance value delivered by planning:

- Documented inspection routines for safety-critical equipment
- Traceable maintenance records for regulatory audits (e.g., GMP, fire safety)
- Accurate labour and material tracking for cost reporting
- Workflow approvals for high-risk work or contractor involvement

Asset Integrity and Lifecycle Transparency

When a planner maintains a clear, up-to-date asset hierarchy and history, they enable long-term thinking: total cost of ownership, lifecycle forecasting, and capital planning. This improves financial governance and allows engineering leaders to make asset decisions based on fact, not gut feel.

Making ESG Part of Daily Planning

The best way to embed ESG in maintenance planning is to treat it as part of the planning quality standard - not a separate initiative. Some practical steps to do this include:

- Add ESG-related questions to your PM audit checklist (e.g., "Is this task contributing to energy efficiency?")
- Track rework or waste incidents and flag as ESG improvement opportunities
- Collaborate with sustainability or ESG officers during major shutdowns or asset upgrades
- Include ESG language in contractor scopes (e.g., expectations for waste disposal or sustainable materials)

Planning with ESG in mind doesn't mean adding extra layers of paperwork or slowing things down. It means planning smarter - recognising that the way we maintain equipment, schedule labour, and consume resources has a broader impact.

Of course, depending on the current condition of your organisation's maintenance plans and execution, this may seem like a luxury angle to take. And probably that's correct. However, in our pursuit for world class maintenance planning, luxury angles like this should absolutely be considered and understood, ready for implementation when the timing is right for your teams.

The planner who sees beyond the work order - who understands that every action has a footprint - becomes a leader in the shift towards sustainable maintenance.

Key Takeaways

- Maintenance planning directly supports ESG performance – By reducing waste, improving energy efficiency, and extending asset life, planners contribute meaningfully to environmental goals.
- Planned maintenance is safer maintenance – Structured planning helps reduce unplanned failures and emergency work, supporting safer work conditions and better compliance with social and safety metrics.
- CMMS data is a hidden ESG asset – When properly structured and maintained, CMMS records can demonstrate regulatory compliance, sustainability efforts, and cost control to stakeholders.
- Energy and emissions should be tracked like reliability KPIs – Planners can integrate energy usage, emissions intensity, or downtime energy cost into job planning and asset reporting.
- Sustainable planning is also lifecycle planning – Avoiding premature replacements, identifying equipment at risk of obsolescence, and promoting repairs over disposal are core planner contributions to long-term sustainability.
- Parts strategy affects waste and cost – Overstocking, obsolete spares, and low-quality replacements all increase environmental impact. Planners help optimise this balance.

Chapter 16

Leadership and Influence

"Who needs a title when you've got trust?"

Maintenance planners often find themselves in a unique and sometimes challenging position within an organisation. They typically operate in a mid-to-senior role but crucially, *without* direct reports. This structure can limit their formal authority, but it opens up the opportunity to lead through influence, collaboration, and credibility. The most effective planners understand that their ability to lead is rooted not in hierarchy but in trust, knowledge, and the ability to bring people together around a shared goal: reliable, efficient maintenance.

Working Across Teams

Planners sit at the intersection of several teams, including:

- **Technicians and Engineers**: These are the front-line workers executing the plans. Planners must speak their language, understand the practicalities of their work, and advocate for the time and resources they need. Often, you'll find it walking a fine line between being 'one of them' but also a bit separated in order to maintain some authoritative influence.
- **Operations/Production**: These are often the planner's internal customers. Operations teams depend on planners to minimise unplanned downtime, optimise planned downtime, and avoid production disruptions. You will also massively want to have them on side for agreeing planned downtimes windows for PM execution.
- **Procurement and Stores**: Planners need to work closely with these teams to ensure parts availability, lead time management, and cost control. We saw in chapter 9 how crucial a well managed spare part strategy is to the maintenance plan.
- **Project Managers and Capital Teams**: As organisations grow and adapt, planners often bridge between ongoing operations and one-off project upgrades or expansions. Being

well connected internally in this area goes a long way to helping you obtain the information you need in a timely manner.

This cross-functional nature of the role requires excellent interpersonal and communication skills. Influence often stems from being seen as a neutral, competent professional who is focused on what is best for the equipment and the business.

Leading from the Middle

Planners are rarely in direct command, but they are often expected to lead discussions, set priorities, and align multiple stakeholders. The concept of "leading from the middle" captures this dynamic well.

To do it effectively, planners must:

- **Build Trust**: Deliver on your commitments. Be consistent and dependable.
- **Demonstrate Expertise**: Know the equipment, the process, the CMMS, and the data better than anyone else.
- **Communicate Clearly and Often**: Keep people in the loop. Be proactive with updates.
- **Act as a Facilitator**: Bring people together to solve problems, rather than dictating solutions.

At Amazon there is a leadership principle of **"Earn Trust"** which is particularly relevant here. In many ways, trust is a planner's most valuable currency. Earned through experience, empathy, and reliability, it allows the planner to influence without a formal title.

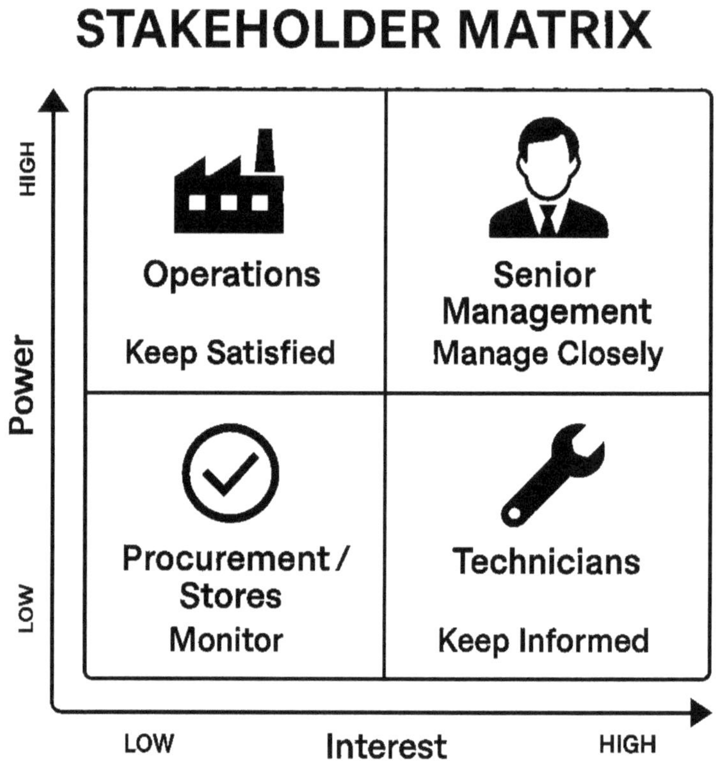

Who Is Your Customer?

This is a deceptively simple question. The reality is:

- **Operations** are your primary customer because their uptime and efficiency are directly impacted by maintenance.
- **Technicians** are also your customers, as you're enabling them to do their job efficiently.
- **Engineering Management** relies on you for reporting, insight, and execution.

- **Procurement and Stores** rely on your input to optimise spend and stock.

The key takeaway: You serve multiple stakeholders. Understanding their perspectives and priorities helps you design plans and strategies that gain broader buy-in and support.

Practical Techniques for Influencing Without Authority

Influencing without direct authority requires deliberate effort. Here are several proven methods planners can use to gain traction:

1. Frame Conversations with "What's in it for them?"

Whether you're talking to operations, technicians, or finance, tailor your message:

- With **operations**, highlight reduced downtime and higher throughput.
- With **technicians**, show how job plans save time and reduce firefighting.
- With **finance**, point to cost avoidance or better asset performance.

Always position your request or suggestion in terms of the listener's priorities as much as you are realistically able to.

2. Use Data as a Neutral Language

Data is persuasive. When recommending a maintenance change or justifying a schedule, bring trend data, CMMS reports, or KPIs. This removes subjectivity and creates a shared ground for decision-making.

For example:

- "This task took 6 hours last time instead of the planned 4. That's been an increasing trend over the last 3 months. We should consider adjusting the job plan or the estimated durations."

3. Storytelling over Statistics Alone

Facts are important, but a brief story or context makes them resonate. Instead of just quoting a 10% increase in PM compliance, try something more like:

> "We've seen fewer breakdowns since we locked in our Monday PM block on the Inbound area.. That one fix last month on Line 2-which avoided a belt failure-saved us four hours of downtime."

It makes the numbers meaningful and in a language your customers are more likely to resonate with.

4. Build Coalitions and Champions

Don't wait until a team meeting to unveil a big change. Build support quietly by:

- Talking to technicians informally and gaining early feedback - test the water
- Checking with operations leaders in advance
- Getting one engineer to help co-author or test a change

Once a few people are on board, the rest tend to follow more easily.

Managing Conflict and Difficult Conversations

As a planner, you may need to challenge priorities, reject low-detail work requests, or push back on unrealistic timelines. Here are some tips to managing those sorts of conversations:

- **Be assertive, not aggressive**: Stand your ground respectfully.
- **Use data to depersonalise the message**: Show trends or task durations to justify changes.
- **Acknowledge the other party's constraints**: Show that you understand their goals.
- **Offer a solution, not just a rejection**: "I can't get this scheduled this week, but here's how we can get it done early next week with full support."

Conflict doesn't have to be adversarial - if handled well, it builds trust.

Personal Branding as a Planner

In a role where authority is often earned rather than given, your personal brand as a planner and as a person is one of your most powerful tools. It's the sum of how you show up-your professionalism, consistency, technical competence, and how others feel after working with you. As a Maintenance Planner, you may not manage people directly, but how you manage information, process, and outcomes shapes how others perceive you, as well as directly affecting their working day.

Remember that perception, in an influence-based role, is everything.

So how can you intentionally build a strong personal brand as a planner?

Be the most prepared person in the room

Preparation earns respect. Whether it's a cross-functional planning meeting, a daily stand-up with technicians, or a project review with leadership-bring the facts, know the schedule, understand the risks, and have options ready. When others see that you consistently do your homework, they'll begin to rely on your input, trust your guidance, and bring you into conversations earlier.

Make results visible, not just effort

A common trap for planners is doing great work behind the scenes-but if no one sees the impact, influence stalls. Find ways to make your contributions visible. Share monthly metrics updates, highlight reduced downtime from improved PM strategies, or demonstrate planning accuracy through digestible reports. Share these through dashboards, brief presentations, or simple team updates. Visibility helps build credibility.

Own the reliability process and CMMS structure

Planners who truly take ownership of the planning process-from work order quality, to CMMS cleanliness, to reliability data-become indispensable. Be the person who understands how your systems work and why they're structured that way. Offer training. Spot and fix gaps in usage. This builds your identity as a trusted process owner and reliability partner, not just an administrator.

Build a reputation for follow-through

People respect planners who do what they say they'll do. It builds confidence and dependability. If you say you'll update a job plan or chase a spare part-do it quickly and confirm when it's done. Over time, this cultivates a personal brand rooted in trust and delivery.

Be solutions-focused, not fault-focused

Maintenance work often uncovers problems-missing spares, poor PMs, reactive pressure. Great planners don’t just report problems; they bring options and solutions. When you consistently help move issues forward, people see you as a problem solver, not just a process gatekeeper.

Engage with curiosity, not criticism

Influence grows when you take time to understand others' perspectives. Ask field technicians why they skipped a step. Sit with procurement to understand lead time issues. Ask operations about their pressures before requesting downtime. Planners who show empathy and curiosity tend to gain more trust than those who jump to blame or judgment.

A strong personal brand doesn't mean being loud or dominant - it means being consistent, capable, and reliable. When others see you as the person who brings order, insight, and improvement to the maintenance process, your influence increases naturally, even without formal authority.

The Influence Index – The KPI You Won’t Find on a Dashboard

In maintenance planning, we obsess (rightfully) over data. PM compliance, wrench time, backlog age - these are our staples. But some of the most important signals of success don’t show up in a CMMS report.

One of the most powerful indicators of a planner’s effectiveness is something we will call “The Influence Index”.

It’s not a graph, it’s a feeling.

- Does operations loop you in on schedule changes?
- How often do people seek your input before they act?
- Do technicians ask for your support or clarification?
- Do managers come to you for insights before project kick-offs?

These are the real-world indicators that you're not just issuing plans - you're shaping the way the maintenance ecosystem works. When field teams trust your plans, when leaders defer to your priorities, when people *want* your input - that's influence. And it compounds.

How to Boost Your Influence Index

You don't need a title to lead. You need consistency, reliability, and trust. Here's how planners can build influence:

- Be the most prepared person in the room – Know the plan, the history, the data. Surprise no one.
- Be proactive – Raise issues before they become problems. Be the one who's always "on it."
- Be responsive – When people reach out, respond quickly and with substance.
- Share success – Publish results and close feedback loops so others see the value of planning.
- Seek input from the field – Before finalising a job plan, ask a technician or team lead for feedback. It shows respect and earns buy-in.
- Stand your ground when it counts – Influence grows when people see you protect quality and reliability, even under pressure.
- Be visible and present – Don't hide behind the desk. Attend stand-ups, walk the floor, be part of the conversation and part of the team.
- Solve problems, not just document them – Bring solutions, not just forms.

Earning Trust with the Field

Perhaps the most overlooked - but most powerful - influence a planner can develop is with the technicians and engineers who execute the work. You're planning the jobs; they're living them. If you lose their trust, your job plans will end up bypassed, rewritten, or ignored.

Here's how to earn that trust:

- Write useful, clear job plans. No fluff, no assumptions. Details that help, not hinder.
- Incorporate their feedback. If they suggest a better method or a tool they always use - update the plan. Let them know you did.
- Show up and listen. Drop in after a job has been executed. Ask what worked and what didn't.
- Be consistent. Follow through on your actions and own your mistakes. Reliability breeds reliability.

When field teams trust your work, everything gets easier. Work is executed more closely to plan. Jobs are safer and smoother. And when something goes wrong, you're the first one they talk to - not the last.

Influence isn't something you ask for - it's something you earn, job by job, person by person. The more trust you build, the more your plans get followed. That's how planners lead without authority.

Mentorship and Informal Leadership

Even without direct reports, planners often mentor others directly or indirectly:

- Help technicians understand the CMMS and job plan process
- Train new engineers or support staff on maintenance workflows
- Provide feedback to shift leads on job performance data

Taking on these informal leadership roles creates goodwill and promotes a culture of continuous improvement.

Key Takeaways

- Planners lead through influence, not authority. Success depends on trust, expertise, and communication-not hierarchy.
- You serve multiple stakeholders. Technicians, operations, procurement, and leadership all rely on your input-so understanding their needs helps drive alignment.
- Effective influence starts with empathy and clarity. Frame conversations in terms of the listener's priorities, and use data and storytelling to support your ideas.
- Conflict is part of the role. Handle it assertively but respectfully, using data to depersonalise challenges and always offering a practical solution.
- Build your personal brand. Be visible, consistent, and reliable. Let your preparation, follow-through, and problem-solving earn you credibility.
- The 'Influence Index' matters. If people regularly seek your input, defer to your planning, and trust your judgement-you're winning, even if it doesn't show up on a dashboard.
- Technician trust is critical. The more your plans help-not hinder-them, the more buy-in you'll get, and the better the work will be executed.
- Mentorship strengthens your role. Even without direct reports, guiding others through CMMS, workflows, or best practices positions you as a leader and builds goodwill.

PART IV – Career and Culture

Chapter 17

Becoming Indispensable

"When people start coming to you first, you're on the right path"

In many maintenance organisations, the planner role is undervalued - seen as admin-heavy or reactive. However, when done well, planning is a strategic capability that can elevate the performance of the entire operation. If you're reading this book, there's a good chance you want to go beyond "just doing the job"-you want to add lasting value, build a reputation for excellence, and become indispensable to your team, site, or company.

This chapter explores ways in which you can make that leap.

How to Stand Out as a Planner

Planners aren't on the tools every day, but we're at the heart of the maintenance engine. If you want to stand out, you need to be more than a work order processor-you need to be a planner with influence, someone who improves reliability, enables efficiency, and helps shape strategic decisions.

Here's how planners can differentiate themselves:

1. Be Proactive, Not Passive

- Don't wait for work to be handed to you. Build forward-looking plans, suggest PM improvements, identify recurring breakdowns before others spot them.
- Actively clean and structure CMMS data without being asked - when others see a well-organised system, they'll attribute it to your diligence. Equally, the opposite will be true.

2. Understand the Business

- Know how the business makes money and how the operations process works. Understand production priorities, customer commitments, and financial constraints.

- When you align maintenance priorities with business objectives, your role becomes more strategic and appreciated by leadership.

3. Speak the Language of Reliability

- Use language like MTBF (Mean Time Between Failures), uptime, asset criticality, and cost-of-failure to communicate in a value-focused way.
- Frame planning wins in reliability terms, not just efficiency (e.g., "Downtime reduced by 20 hours this quarter due to improved task planning").

4. Help Others Succeed

- Build plans that make life easier for technicians. Clarify scopes for operations. Support procurement by forecasting needs.
- When other departments see you as an enabler, not a bottleneck, your value rises significantly.

5. Be Consistently Reliable

- Be the person who never drops the ball. Trusted, accurate, predictable.
- Deliver clean job plans on time, with everything the team needs to execute. That reliability builds your own credibility.

Career Development and Certification

Becoming indispensable isn't just about how you show up-it's also about how you grow your skills and validate your knowledge.

Industry Certifications Worth Considering:

- **SMRP / CMRP (Certified Maintenance & Reliability Professional)**
    - Offered by the Society for Maintenance & Reliability Professionals.
    - Globally recognised and focused specifically on best practice maintenance planning and reliability.
    - Excellent for mid-career planners looking to validate their professional capability and improve career mobility.
- **IAM Certificate/Diploma and ISO 55000**
    - These focus more on asset management strategy than tactical maintenance.
    - Ideal if you want to progress into leadership roles, capital planning, or cross-functional asset management roles.
- **Apprenticeship Completion or Technical Trade Certificates**
    - These remain valuable foundational credentials. For planners who came from the tools, showcasing your trade background alongside planning acumen is a powerful combination.
- **Internal Training and Leadership Courses**
    - Many large organisations offer structured development pathways. Don’t overlook them-they show initiative and help you build cross-functional relationships so understand the options available to you.

Integrating Planning into the Organisation

To truly become indispensable, you must embed planning into the fabric of how the organisation runs. This means pushing past the idea of planning as just a maintenance function, and helping others understand how your work impacts their success too.

How to Do This:

- **Operations**:
    - Make planning part of the daily and weekly operations rhythm.
    - Use planning boards, joint planning meetings, and shared KPIs (e.g. planned downtime compliance) to align better with production goals.
- **Procurement and Stores**:
    - Use planning forecasts to influence purchasing decisions.
    - Keep procurement in the loop early to avoid expedited shipping costs or stock-outs.
- **Finance**:
    - Link your planning to maintenance spend visibility.
    - Demonstrate that accurate job plans help control labour costs, optimise spare usage, and avoid expensive reactive failures.
- **Engineering and Projects**:
    - Help capital project teams by contributing asset metadata, lifecycle strategies, and long-term maintenance planning for new equipment.
    - Ensure new assets are fully structured into the CMMS and maintainable from day one.

Becoming the "Go-To" Planner

What does it mean to be indispensable? It means:

- When there's a critical shutdown, people ask for *you* to do the planning.
- When the site's backlog grows out of control, leadership trusts *you* to bring it back into balance.
- When auditors or senior managers need to understand the maintenance strategy, *you* are called to present it.

That doesn't happen by accident. It happens when you consistently bring value, grow your expertise, and collaborate across boundaries.

Here are a few final habits of indispensable planners:

- **Be a student of your craft.** Read, network, learn, improve.
- **Keep your CMMS clean and your job plans clear.** These are your tools and your signature.
- **Be calm under pressure.** Planners often act as a bridge between chaos and order.
- **Never stop asking: "How can I make this better?"**

By doing all this, you don't just survive in the planning role - you thrive. You become the backbone of the maintenance function, and a valued partner in the organisation's success.

And that's the true meaning of being indispensable.

Key Takeaways

- Don't just do the job-own the role. Great planners go beyond processing work orders; they proactively drive reliability, efficiency, and cross-functional collaboration.
- Be proactive, not reactive. Clean data, forward-looking plans, and recurring failure analysis set top planners apart.
- Understand the bigger picture. Know how your organisation runs, where value is created, and how maintenance supports that value.
- Use the language of reliability. Communicate wins using terms like MTBF, uptime, and cost avoidance to resonate with leadership.
- Support the success of others. Make life easier for technicians, operations, procurement, and leadership-this builds influence and trust.
- Consistency builds credibility. Deliver clean, complete, timely plans-every time. People remember the planner they can rely on.
- Invest in your development. Certifications like CMRP or IAM, and internal training programs, signal commitment and expand your opportunities.
- Embed planning in the organisation. Collaborate with operations, procurement, finance, and capital projects to make planning a core function, not a siloed task.
- Be the "go-to" planner. The one trusted for critical shutdowns, backlog recovery, or audits. Earn that trust by adding consistent, visible value.
- Strive for continuous improvement. Whether it's your CMMS, your job plans, or your own skills-keep asking: "How can I make this better?"

Chapter 18

Planning Culture

"Culture is built in the routines - and reinforced in the small details"

Creating a great maintenance plan is one thing. But creating a planning culture-a way of working where structured maintenance is prioritised, respected, and supported across all levels of the organisation-is another challenge entirely.

In many operations, planners find themselves fighting an uphill battle. Maintenance is often seen as reactive by nature. Leadership is focused on immediate output. And technicians are pulled in every direction by urgent breakdowns.

But if you want to truly drive long-term reliability, cost control, and team performance, culture matters.

This chapter explores how to build a culture where maintenance planning is embedded into the fabric of the operation-not as an administrative function, but as a strategic enabler.

Building Reliability-Minded Teams

A planning culture starts with mindset. Everyone, from leadership to technicians, needs to value reliability and structured execution over urgency and improvisation.

Here's how to start shifting the team's mindset:

- **Talk about failures as opportunities to learn, not just things to fix.**
  After a breakdown, don't just celebrate the fast response. Celebrate the analysis, the root cause identification, and the long-term fix.
- **Involve technicians in planning feedback loops.**
  When frontline teams see that their feedback on job plans and PMs is acted on, they're more likely to contribute thoughtfully and follow structured plans.

- **Balance speed with quality.**
  Set expectations that a job done right the first time is more valuable than rushing to close a ticket. This is a key cultural shift.
- **Recognise and reward proactive behaviours.**
  Celebrate when teams execute a planned job exactly as written, or when a PM catches a failure before it happens.

You can't build a planning culture without the field team on your side. They don't have to love every spreadsheet - but if they trust that your plans make their lives easier and safer, that's the start of culture change.

Educating Leadership on Planning's Value

For a planning culture to stick, it needs top-down support. Leaders have to understand what planners do, how planning impacts uptime and cost, and why unplanned work is the silent killer of operational efficiency.

Here's how planners can educate and influence leadership:

- **Translate technical wins into business impact.**
  Don't just report that "50% of jobs were planned." Show how that reduced overtime costs, or improved equipment availability.
- **Bring visibility to the cost of reactive work.**
  Break down the hidden impacts: expedited parts, overtime, missed production windows, safety exposure.
- **Share planning metrics in business reviews.**
  Integrate KPIs like PM compliance, backlog age, or schedule adherence into leadership dashboards.
- **Invite leadership to planning meetings or field walkdowns.**

Seeing the process builds appreciation. Explain how weeks of coordination go into a single well-executed PM shutdown.

- **Position planning as risk management.**
  Planning isn't just about paperwork. It's about preventing failures that could impact customer delivery, compliance, or safety.

ISO 55000 and Asset Management Frameworks

A true planning culture doesn't exist in a vacuum - it often reflects how well the organisation understands asset management as a whole. That's where international standards like ISO 55000 can help.

What Is ISO 55000?

ISO 55000 is a global standard for asset management. It provides a framework for managing the value of physical assets across their lifecycle-not just their operation or maintenance.

It's not a planning manual, but it sets out clear principles for:

- Aligning asset decisions with business goals
- Managing risk and performance systematically
- Ensuring clear roles and responsibilities
- Supporting long-term, data-informed decisions

Why Is It Useful for Maintenance Planners?

Planners live in the operational layer of asset management. ISO 55000 shows that planning isn't just about scheduling-it's part of a wider system that connects:

- Strategic business objectives

- Asset lifecycle management
- Financial forecasting
- Reliability and risk analysis

Understanding ISO 55000 principles helps planners:

- Speak the language of senior management
- Align their work with long-term asset plans
- Justify investments in better tools, training, and planning time

How Should Planners Apply ISO 55000 Guidance?

You don't need to be an ISO expert. But here's how you can apply the mindset:

- Think in terms of asset lifecycle, not just next week's jobs
- Push for accurate asset hierarchies and metadata in the CMMS
- Make sure each job plan supports broader reliability goals
- Ask: "Does this job contribute to our strategic asset plan?"
- Work closely with asset owners, engineers, and finance to link maintenance to long-term value

Creating Top-Down Line of Sight

In a true planning culture, people at every level understand how their actions tie into broader business outcomes. This is called line of sight-and planners play a key role in making it happen.

What Line of Sight Looks Like:

Level	Focus	Example
Executive	Strategic goals (e.g. uptime, cost per unit, compliance)	“We want to reduce downtime by 15%”
Maintenance Manager	Resource planning, high-level KPIs	“Let’s improve PM completion rate to reduce breakdowns”
Planner	Job-level execution, schedule compliance	“I’ll build a 4-week plan to ensure 90% PM compliance”
Technician	On-the-ground task execution	“I’m executing a well-planned PM that prevents a costly failure”

Planners can bridge the gap between these levels by:

- Making reliability data visible and relevant to leadership
- Ensuring technicians understand how their jobs support larger goals
- Using planning reports to show trends, wins, and gaps

When planning is clearly connected to business success, it gains traction, visibility, and respect.

Building a planning culture isn’t just about processes or CMMS usage, those are just the foundations. It’s about changing how people think and behave, from the technician to the CEO.

Reinforcing Planning Culture Through Rituals

Once you've laid the groundwork for a strong planning culture, it must be reinforced daily through team rituals, communication cycles, and shared behaviours. Culture isn't set by slogans-it's shaped by what people consistently see and do.

Key routines that support strong planning culture:

1. Weekly Planning Meetings

- Align maintenance, operations, and procurement.
- Review the schedule for the upcoming week.
- Confirm parts, access, permits, and resources.
- Adjust for known absences or priority shifts.

Planner's Role: Lead the meeting with clarity, keep discussions productive, and document changes.

2. Daily Stand-Ups

- Quick 10–15 minute check-ins with maintenance and ops teams.
- Review yesterday's completions and today's plan.
- Address urgent blockers and communicate schedule changes.

Planner's Role: Stay close to the floor - listen to issues, record feedback, and feed it back into future plans.

3. Monthly KPI Reviews

- Present planning metrics to leadership and frontline teams.
- Discuss trends: PM compliance, backlog, plan accuracy, wrench time.
- Use the session to educate and highlight the value of structured planning.

Planner's Role: Translate data into insights and suggest clear next steps.

4. Post-Job Debriefs (Selective)

- For complex or high-impact jobs, run a quick debrief with the field team.
- What went well? What could be improved in the plan?
- Capture these for job plan refinement.

Planner's Role: Lead with curiosity, not criticism. Show that their feedback gets actioned.

These routines help to turn planning from a theoretical concept into a living practice. They keep everyone aligned, build trust, and give planners a regular voice in team operations.

Overcoming Resistance to Structured Planning

In reality, not everyone loves structure. In reactive organisational cultures, a lot of the planning routines we've explored can feel like bureaucracy. You might hear:

> "We don't have time to plan, we just need to get it fixed."
> "The plan never matches the real job anyway."
> "We've always done it this way."

So if you walk into an environment like that, how do you begin to shift the mindset?

Tactics for handling resistance:

1. Start with Wins

- Use small pilot areas or teams to demonstrate planning success.

- Track before/after metrics: fewer breakdowns, less overtime, faster turnaround.

"Look what we achieved in Line 2 last month with a new planning process. Imagine that concept rolled out across the entire site."

2. Involve the Frontline

- Don't write plans in isolation.
- Co-create job plans with experienced technicians.
- Ask for feedback, implement it, and close the loop.

3. Solve Real Problems

- Don't push a rigid system. Tailor the planning process to *actually solve* pain points-like missed parts, poor communication, or overbooking.

"This planning board isn't for compliance. It's here so you don't walk halfway across the plant for a missing gasket."

4. Celebrate and Recognise

- Recognise the tech who followed a great plan and provided helpful feedback.
- Acknowledge the team that cleared their weekly plan 100%.

5. Stay Visible and Supportive

- Be on the floor. Be responsive. Fix bad plans fast.
- Show that planning is a support function, not an ivory tower.

Culture change is long-game work-but planners are perfectly positioned to drive it. Why? Because planners touch *everything*:

people, process, parts, data, and downtime. With visibility and influence across the operation, planners can be the cultural architects who turn fire-fighting into forward-thinking.

By embedding rituals, defining clear standards, and navigating resistance with empathy and evidence, planning evolves from a process into a culture-and that's where real reliability lives.

Key Takeaways

- Culture matters as much as process. A strong planning culture ensures structured maintenance is respected, prioritised, and embedded across the operation-from technicians to leadership.
- Planners help shift mindsets. By involving technicians, celebrating proactive wins, and treating failures as learning opportunities, planners can lead a change in how maintenance is perceived.
- Educate leadership on planning's impact. Use business-focused language, data, and visibility to show how structured planning drives uptime, reduces costs, and manages risk.
- Adopt an asset management mindset. Align your planning work with ISO 55000 principles-think in terms of lifecycle value, not just next week's jobs.
- Create line of sight. Help connect technician tasks to planner goals, manager KPIs, and executive strategy-making everyone part of the reliability story.
- Reinforce culture with consistent rituals. Weekly planning meetings, daily stand-ups, monthly KPI reviews, and post-job debriefs all help make planning real and repeatable.
- Overcome resistance with empathy and results. Start with small wins, involve the frontline, and tailor the process to solve actual pain points-not just enforce compliance.
- Planners are culture builders. Your role touches every part of the maintenance ecosystem. Use that visibility to drive a shift from reactive chaos to reliable, structured execution.

Chapter 19

The First 90 Days In a New Planner Role

"Listen first. Fix second. Deliver early wins - then build your system"

Laying Foundations, Building Influence

Starting a new job as a maintenance planner is exciting - and daunting. Whether it's your first planning role or just a new environment, those first 90 days represent a crucial window of opportunity. It's when you establish your reputation, shape how others perceive your value, and begin to influence the systems and people around you.

But here's the challenge: no two organisations are the same.

You might be stepping into a company with mature systems and clear processes, or conversely, into one where chaos is the norm and you're the first planner they've ever hired (I definitely prefer the latter!). The CMMS might be world-class… or a neglected mess. The team might welcome you with open arms… or treat you with caution until you prove yourself.

Because of all this, your mindset really tends to matter more than checklists in your early days.

Think Like a Consultant, Act Like a Team Member

The best planners enter new roles with a dual mindset. First, they act like a consultant - observing systems, listening to frustrations, mapping pain points, and looking for opportunities. But at the same time, they act like a team member - building trust, contributing visibly, and showing humility. You're not there to criticise. You're there to understand, improve, and make everyone's job easier.

Play the Long Game, Win Early Rounds

One of the most common mistakes new planners make is trying to fix everything at once. You'll likely encounter broken processes, confusing job plans, parts issues, and data gaps. But change takes time - and influence must be earned.

Focus on small, strategic wins in the first few months. Look for low-hanging fruit that solves a real problem. Use those early successes to demonstrate your approach, gain confidence from the team, and earn permission to tackle bigger changes later.

Relationships Over Systems

No matter how good your planning knowledge or CMMS skills are, your ability to make an impact will come down to one thing: relationships. The technicians, supervisors, operators, and procurement teams - these are the people who will make or break your plans. In the first 90 days, every conversation is an opportunity to build credibility.

Be present. Be respectful. Ask questions with genuine curiosity. Show that you're not just there to “process work orders”; you're there to help the team succeed.

Own Your Onboarding

Most companies have some form of onboarding - but very few truly onboard planners - part of the reason for the creation of The Planner HQ, and this very book. This minimal onboarding means you need to be proactive. Design your own discovery process. Map out what you need to learn, who you need to meet, and how you'll build your understanding of the site, systems, and culture. Don't wait for clarity to be handed to you - be curious and go find it.

Define Your Own Success

Every planner role is a little different. You might be expected to own the full planning cycle - or just take lead on job plans and backlog. You might be tasked with improving the CMMS, or building it from scratch. Take time early on to understand what success looks like - not just in your eyes, but in the eyes of your manager, the team, the business, and your new customers.

Then, build your plan to get there.

This chapter gives you a roadmap for those crucial early days. But more importantly, it should empower you with a mindset. The job you walk into will have its flaws. Every team does. But you're not just a new hire. You're a change agent - a builder of structure, trust, and reliability.

And it all starts now.

Week 1–4: Listening and Learning

Your first month should be about observation, orientation, and relationship-building. Resist the temptation to dive in and "fix" things. You'll be far more effective if you understand the landscape first.

Your goal is to build context - the systems, the people, the history, and the pain points.

Focus Areas:

- CMMS and Process Understanding
    - Explore the system hierarchy, asset structure, work order lifecycle, and reporting tools.
    - Review how job plans are stored, who raises work requests, and how approvals flow.
- Walk the Floor
    - Spend time with technicians, engineers, and supervisors on the shop floor.

- Ask questions, observe job execution, and get a feel for site culture.

- Stakeholder Introductions
    - Schedule short introductions with operations leads, planners (if any), procurement, and stores.
    - Find out what their expectations are of the planning function.

- Document Review
    - Skim through existing PMs, backlog work orders, and job plan libraries.
    - Spot inconsistencies, duplications, or signs of neglect.

- Ask the Magic Question:

  "What's working well in our planning process - and what's broken?"

This first phase is all about earning respect by listening. Show curiosity. Take notes. Ask why. Be visible, engaged, and humble. Trust is earned by how you show up, not what you know.

Week 5–8: Early Wins

Once you've gained insight, it's time to act - but carefully. Don't try to overhaul the system. Instead, pick 1–2 high-friction areas where you can make a visible, meaningful improvement. The key is to build momentum and credibility.

Target areas for early wins:

- Backlog Clean-Up
  - Triage a small area of backlog - maybe a single department or asset group.
  - Close obsolete jobs, clarify scopes, and schedule what's doable.
- Fix a PM
  - Choose a PM with technician complaints or poor effectiveness.
  - Rewrite it clearly, add missing steps or parts, and test the updated version.
- Solve a Part Problem
  - Identify a job that was recently delayed by a missing or misidentified part.
  - Work with stores to fix the issue, and update the job plan or BOM accordingly.
- Share Insight
  - Produce a short report showing a useful trend - e.g., the top 5 causes of reactive work.
  - Share it with leadership, ideally paired with a simple proposed action.

- Plan and Align Downtime
  - Coordinate a small downtime window with operations and execute 100% of scheduled work.
  - Use this as a case study to show the value of planning.

Your early wins should demonstrate competence, responsiveness, and partnership. Don't go big - go smart. Prove that your work makes people's lives easier and operations more efficient.

Week 9–12: Building Structure

With early credibility established, the final phase of your 90-day plan is about laying the foundation for scalable, repeatable planning practices.

This is where you begin shifting from reactive contributor to process owner.

Key moves to consider:

- Start Weekly Planning Meetings
  - If not already in place, create a simple structure with maintenance, operations, and stores.
  - Keep it tight - review backlog, upcoming PMs, parts readiness, and workforce availability.
- Develop a Standard Job Plan Library

  - Choose a handful of recurring jobs and write clear, reusable plans for them.
  - Tag them appropriately in the CMMS for easy future access.

- Build a Personal Dashboard
  - Identify a few KPIs that matter - backlog age, schedule compliance, planner accuracy.
  - Track them monthly to measure impact and share progress.

- Create Feedback Loops
  - Add a feedback field to completed work orders or run monthly debriefs with technicians.
  - Show that input leads to action by updating job plans and closing the loop.

By now, you should be embedding yourself into the system - not just working within it. Your actions should be driving structure, visibility, and repeatable quality.

Earning Trust Early

Across all three phases, your most valuable currency is trust. Without it, plans are ignored, feedback is withheld, and your influence shrinks. With it, you become indispensable.

Ways to earn trust fast:

- Be the most prepared person in the room
- Do what you say you'll do - every time
- Ask thoughtful questions that show you understand both people and process
- Follow up when you don't know something
- Give credit to others when things go well

Trust builds slowly, but it compounds.

Setting a 6–12 Month Vision

At the 90-day mark, pause and look ahead. You've gathered insight. You've fixed some small things. You've begun to add structure.

Now ask: Where do I go from here?

Start drafting a personal roadmap. Think in two tracks - strategic and tactical.

Strategic Thinking:

- What major planning gaps exist?
- Are PMs effective and aligned to failure modes?
- Is CMMS data usable and accurate?
- Where is most of the reactivity coming from?

Tactical Goals:

- Create 50 standard job plans
- Review and optimise top 10 PMs
- Launch a planner performance dashboard
- Complete CMMS or reliability training

Document your vision, even if informally, and share parts of it with your manager. It shows you're not just doing the job, you're owning it.

Your first 90 days won't make or break your career - but they will set the tone. How you show up, what you prioritise, and how you communicate during this window shapes how people perceive your value.

If you can consistently:

- Listen deeply
- Act thoughtfully
- Solve real problems
- Communicate clearly
- Build trust intentionally

...you won't just survive the transition. You'll thrive in it, and lay the foundation to become a transformational planner within your new team.

Planner's Toolkit: Download Your First 90 Days Checklist

Turn this chapter into action with a free downloadable version of the 90-Day Planner Checklist - ready formatted for easy tracking, with clickable checkboxes and space for your notes so you can hit the ground running in your first 90 days!

✅ Keep yourself on track

✅ Capture insights as you go

✅ Build momentum in your new role

Get it free at **www.theplannerhq.com**

Just drop your email and we'll send it straight to your inbox.

DOWNLOAD THE

FIRST 90 DAYS PLANNER CHECKLIST

Visit www.theplannerhq.com to access this free resource

Chapter 20

BONUS CHAPTER

Planner FAQ's

"The best planners ask better questions - and never stop learning"

Over the course of this book, we've covered a wide range of principles, strategies, and best practices for effective maintenance planning. But even the best-structured guide can leave questions hanging - especially in a role as complex and varied as the Maintenance Planner.

That's where this FAQ section comes in.

We've compiled a set of common - and sometimes critical - questions that planners regularly ask in the field, in interviews, during training sessions, or when stepping into new responsibilities. These questions are designed to be practical, direct, and rooted in the real-world demands of the role.

Whether you're just starting out, mentoring others, or simply want to validate your current practices, this section offers a quick-access knowledge base that complements the deeper concepts explored throughout the book. Keep it handy as a reference - and share it with others on the planning journey.

1. What's the difference between a planner and a scheduler?

While the roles are often combined or confused, there's a clear distinction between planning and scheduling. A planner focuses on how the job will be done - defining the scope of work, ensuring the right parts and tools are available, writing the job plan, estimating time, and confirming asset readiness. The scheduler, on the other hand, is concerned with when the work will be done - assigning tasks to resources, managing time windows, and balancing priorities across the team. Think of the planner as preparing the playbook and the scheduler as calling the plays on game day. In smaller teams, these roles are often handled by one person. In larger or more mature organisations, they're separated to allow more focus and efficiency in each discipline.

2. How do I prioritise jobs when everything feels urgent?

When everything is marked as urgent, nothing truly is. Effective job prioritisation starts with a clear understanding of asset criticality - which equipment has the greatest impact on safety, production, cost, or compliance. Establish a priority matrix that weighs consequence of failure against urgency, and socialise it across operations and maintenance. Challenge vague or high-priority requests without supporting rationale, and use data to reinforce your case - has the job been sitting idle for two weeks? Is there a PM already scheduled soon? Engage with requestors to understand true impact. When pressure builds, planners must bring order. Stick to your process, communicate openly, and be prepared to explain how priorities were set. Long-term trust builds when others see a consistent, fair prioritisation method at work.

3. What's the best way to clean up a messy backlog?

A disorganised backlog clogs planning workflows and hides real priorities. Start by exporting your backlog data from the CMMS and categorising it by age, asset, and type of work. Archive or cancel jobs that are clearly obsolete, duplicated, or already completed. Next, identify "quick wins" - jobs that can be bundled together or completed with minimal prep. Engage the field team to validate open work: are the jobs still needed? Have conditions changed? Are parts still available? Build a structured backlog review process into your weekly or monthly rhythm so cleanup becomes ongoing, not a once-a-year fire drill. Over time, aim to segment the backlog into manageable buckets - planned, awaiting parts, waiting schedule, waiting access - to bring transparency and control to the process.

4. How should I structure an equipment hierarchy in the CMMS?

A well-structured equipment hierarchy is foundational to effective planning. It should mirror the physical layout and functional relationships of your plant or site. A typical structure might flow from Facility → Area → System → Equipment → Component. Each level helps you understand context, group related tasks, and enable smarter reporting. Avoid overcomplicating with too many layers, but do aim for consistency and clarity. Use naming conventions that are meaningful to both technicians and planners. A consistent hierarchy also supports more accurate failure analysis, job history tracking, and spare part linking. Planners should work with engineering and reliability teams to periodically review and improve the structure - it's not just a database field, it's a vital planning tool.

5. What does 'bad data in = bad data out' really mean for planners?

This phrase is more than a cliché - it's a warning. Planners rely heavily on historical data to refine job plans, set frequencies, measure compliance, and report performance. If work orders are closed without meaningful notes, if failure codes are inconsistent, or if time estimates are inaccurate, the CMMS becomes a misleading tool. Poor data leads to poor decisions: over-maintenance, missed failure patterns, misallocated resources. Planners must champion data quality by setting standards, training technicians, and auditing completed work orders. That doesn't mean chasing perfection - it means focusing on consistency, clarity, and relevance. High-quality data transforms the CMMS from a digital filing cabinet into a planning powerhouse. Your job as planner isn't just to use data - it's to cultivate it.

6. What's the role of a planner in Root Cause Analysis (RCA)?

Planners may not always lead formal RCA investigations, but their involvement is critical. You have visibility of job history, technician notes, part replacements, and failure patterns - all of which are essential inputs into root cause analysis. You can support by pulling complete work order histories, highlighting recurring issues, and identifying gaps in PMs or job plans that may have allowed the failure to develop. More importantly, once an RCA is complete, planners are the ones who close the loop. That means translating RCA findings into updated job steps, new inspection points, adjusted frequencies, or training requirements. The real value of RCA is in the changes it drives - and planners are uniquely placed to make those changes stick.

7. How do I build trust with technicians?

Trust is earned through consistency, clarity, and respect. Write job plans that are clear, helpful, and realistic - avoid generic copy-paste steps. Ask for feedback, and act on it when it makes sense. Be visible: attend stand-ups, walk the floor, ask questions. Don't pretend to know everything - if you don't understand something, say so and learn from the experts. Follow through on your commitments - if you say you'll get a part, investigate an issue, or update a plan, do it. Over time, field teams will see you as someone who supports their work, not just adds process. When that trust is in place, you'll get better information, smoother execution, and stronger collaboration - all of which make planning more effective.

8. What should I do when a job plan doesn't go as expected?

Every deviation is an opportunity to learn. First, talk to the technician - what actually happened, and why did things unfold differently than planned? Was a step unclear? Were tools missing? Did they find something unexpected? Review the work order closure notes and any time discrepancies. Update the job plan immediately if the task will be repeated. If it was a one-off issue, document the lesson learned anyway - it may resurface later. Consider whether this case reveals a broader pattern (e.g. similar jobs often running over time) and use that insight to refine future planning. Great planners don't treat deviations as failures - they treat them as feedback. The goal is not perfection, it's continuous improvement.

9. How can I manage planned downtime more effectively?

Start by treating planned downtime like a project. That means building a scope of work, identifying dependencies, confirming materials, and locking in labour early. Work backwards from the target window: what tasks can realistically be done in that time? What can be grouped? Use the CMMS to reserve or kit parts, and confirm tool availability. Communicate early with operations to agree on the timing, access, and any pre-shutdown prep. Use visual scheduling tools (like a Gantt chart or wall planner) to map the work and coordinate with other stakeholders like contractors or production planners. During execution, monitor progress and resolve issues fast. Afterward, debrief - what worked, what didn't, and what should be adjusted next time? Strong planned downtime performance is one of the planner's clearest KPIs.

10. What KPIs should I be tracking as a planner?

Your KPIs should reflect both planning process health and maintenance outcomes. Key metrics include:

- Schedule Compliance – What % of the weekly schedule was completed?
- PM Completion Rate – Are preventive tasks being done on time?
- Reactive vs. Planned Work Ratio – Is your program proactive or fire-fighting?
- Backlog Age – How old are your open work orders?
- Job Plan Accuracy – Is estimated vs. actual time within an acceptable margin?
- Wrench Time – Are technicians spending most of their day doing value-adding work?

Just as importantly, look at data quality indicators - are work orders being closed with good notes? Are failure codes being applied? Don't chase metrics for vanity - use them to drive decisions, justify improvements, and demonstrate the value of good planning.

11. How detailed should a job plan be?

The right level of detail depends on the complexity of the task, the experience of your technicians, and the criticality of the equipment. For high-risk or infrequently performed work, include clear step-by-step instructions, tools and materials, safety precautions, and acceptance criteria. For simpler or well-understood tasks, a shorter plan may suffice - but still include key checks, part numbers, and estimated time. The rule of thumb is this: Could a competent technician unfamiliar with this job complete it safely and successfully using only this plan? If not, it needs more detail. Don't overwhelm with filler - focus on clarity, sequencing, and the essentials. A job plan is not just an instruction sheet - it's a tool for standardising quality and reducing variability.

12. What's the difference between a job plan and a work order?

A job plan is the template - a predefined set of instructions, tools, parts, and time estimates for a specific task. It exists outside of time, as a reusable resource. A work order is the live instance of work to be performed - it references the job plan, but includes context like the date, assigned technician, asset location, and priority. When a planner schedules a PM, the CMMS usually pulls in the job plan to populate the work order. But the work order becomes the living record - capturing what actually happened, how long it took, and any deviations. Both are critical: job plans drive consistency; work orders document reality.

13. How do I get operations to support the maintenance plan?

Start by speaking their language - availability, throughput, risk, cost. Don't just say "we need downtime" - say, "this task reduces the chance of a 6-hour breakdown during your peak shift." Involve them early in weekly planning meetings. Show them the win-win: how planned work avoids production disruption. Track and share metrics - how many hours of unplanned downtime you've avoided this month through proper planning. Build trust by being consistent and responsive. When conflicts arise, offer options - "we can move this to Friday, but it increases failure risk." Over time, a good planner becomes not just a requester of downtime, but a partner in productivity.

14. How do I handle incomplete or low-quality work requests?

Poor requests are a productivity killer - they clog your backlog and create confusion. Build a clear work request standard: what information must be included (asset, problem description, urgency, requestor). If a request doesn't meet the standard, return it with a polite prompt for more detail - ideally via the CMMS so you maintain an audit trail. Work with supervisors to coach their teams on how to raise usable requests. Involve technicians too - their feedback can help identify repeat offenders or request types. Over time, you can reduce noise in your backlog and focus on meaningful, actionable work. Quality in → quality out.

15. How should I prioritise work in the backlog?

Prioritisation isn't just about severity - it's about risk, impact, and timing. Use a structured framework that considers:

- Asset criticality – What happens if this fails?
- Failure consequence – Is it safety, compliance, quality, or just nuisance?
- Lead time – Are parts or permits required?
- Timing – Is there a window where the work can be done with least disruption?

  Assign a priority level that aligns with your organisation's planning cadence. For example:

- Priority 1 – Immediate (emergency)
- Priority 2 – Plan this week
- Priority 3 – Plan this month

- Priority 4 – Low priority / bundle later

Revisit your backlog weekly. A low-priority job today might become urgent if the condition deteriorates. Good planners treat backlog like a living portfolio - actively managed, not ignored.

16. How often should I review or audit my PMs?

A good rule of thumb is to review each PM at least once per year - but the real answer depends on asset criticality and performance. High-risk or high-cost assets may need quarterly reviews. Trigger reviews after key events too: repeated failures, missed inspections, or technician complaints. Don’t audit just for the sake of it - aim to improve clarity, accuracy, and relevance. Involve technicians during reviews and always ask: Does this task still target a real failure mode? Are the steps clear? Is the frequency appropriate? Build a formal PM audit schedule if possible and track version control in your CMMS so you know when a task was last updated.

17. What’s the best way to start improving our CMMS data quality?

Start small, start structured. Pick one area - such as completed work orders or asset metadata - and begin auditing. Look for missing fields, inconsistent terminology, duplicate records, or vague descriptions. Clean-up is most effective when paired with user training: make sure technicians know what “good” looks like. Set standards (naming conventions, dropdown lists, required fields) and reinforce them with spot checks and feedback. Consider a “data quality dashboard” to monitor trends. Most importantly, don’t try to fix everything at once.

Continuous improvement wins over one-time cleanups. Make it a habit - not a project.

18. Should I standardise job plans across similar assets?

Yes - but with caution. Standardising job plans saves time, improves consistency, and helps you scale. For example, you might have a generic PM for "Conveyor – Monthly Inspection" applied across ten lines. But make sure each instance is reviewed against the specific context: age, model, operating environment, known failure modes. Add asset-specific notes where needed, or create child job plans linked to a master template. Avoid blindly cloning plans - they can drift out of relevance fast. Treat standardisation as a tool for efficiency, not a substitute for asset understanding. Good standardisation still leaves room for nuance.

19. How can I make my weekly planning meetings more effective?

Preparation is key. Go in with:

- A clear draft schedule
- A list of jobs that need signoff or clarification
- Visibility of parts availability and technician capacity

Keep the meeting tight and structured: review last week's performance briefly, confirm this week's plan, and resolve blockers. Assign actions and document changes - don't leave agreements verbal. Invite the right people (maintenance, operations, maybe procurement if part-sensitive). Use visual tools - a planning board or shared screen helps focus discussion. If meetings feel repetitive or chaotic, review your agenda and tighten scope. Remember: the meeting is the heartbeat of your week - not a formality.

20. What's the best way to build credibility quickly in a new planner role?

Focus on listening, learning, and delivering small wins early. Spend your first few weeks walking the floor, talking to technicians, reviewing the CMMS, and asking smart questions. Don't rush to "fix" everything. Instead, identify one or two quick wins - cleaning up a PM, fixing a spare part issue, improving a recurring job plan. Be visible, follow through on your promises, and avoid pretending to know things you don't. Show curiosity, not ego. When field teams and managers see that you care, understand their world, and deliver - trust builds. And from trust comes influence.

21. How can I improve technician engagement with my job plans?

Technician buy-in is critical to the success of any maintenance plan. To improve engagement, involve technicians early in the process by seeking their feedback on job plan clarity, task sequencing, and common pain points. Incorporate their suggestions into the job plans and let them know when updates are made based on their input. Keep plans practical, concise, and free from unnecessary filler - each step should add value. Avoid assumptions about what technicians already know, and don't overload tasks with administrative requests. Where possible, provide supporting visuals, such as diagrams or photos, especially for complex or infrequent jobs. Most importantly, follow up: ask how the plan worked in practice, and use the feedback loop to refine future plans. When technicians see their input driving real change, they become more invested in the planning process.

22. What's the difference between a PM and a PdM task in the CMMS?

Preventive Maintenance (PM) tasks are scheduled based on fixed intervals - such as time, usage hours, or calendar dates - with the goal of preventing equipment failure through routine checks, lubrication, adjustments, or part replacement. Predictive Maintenance (PdM), on the other hand, is condition-based and relies on data to determine when maintenance is actually needed. PdM tasks are typically triggered by sensor data (like vibration, temperature, or oil analysis) that indicate equipment degradation. In the CMMS, PM tasks usually appear as recurring work orders with fixed frequencies, while PdM tasks may be triggered manually or automatically based on asset condition thresholds. Planners must understand the distinction because it affects how jobs are scheduled, justified, and prioritised - and how the organisation balances cost vs. reliability.

23. How should I handle urgent reactive work that disrupts the plan?

Reactive work is inevitable in any maintenance environment - but how you manage it determines whether it becomes a chronic issue or a short-term interruption. First, assess the urgency: is the work truly critical, or can it be scheduled at the next available window? If it must be done immediately, communicate the impact on the existing schedule to relevant stakeholders (e.g., operations, technicians). After the fact, make sure reactive jobs are closed properly in the CMMS with accurate notes and failure data. During weekly planning reviews, track how often reactive work disrupts the schedule and look for root causes. Were there warning signs? Was it preventable? Feed insights into job plans or PM reviews to avoid repeat disruptions. Planners can't eliminate all emergencies - but they can turn them into learning opportunities that strengthen the plan.

24. What's the best way to approach critical spares reviews?

Critical spares reviews are essential to ensuring the right parts are stocked for the most important assets. Start by identifying critical equipment (based on safety, production impact, or cost of failure), then map the components most likely to fail or require replacement. Use historical data to identify parts used during past failures and confirm current stock levels. Involve maintenance technicians, reliability engineers, and procurement in the process to get a rounded view. For each part, assess lead time, cost, shelf life, and likelihood of failure - and weigh those factors against stocking cost. Document rationale for each decision, and set appropriate min/max levels in the CMMS. This review should be repeated annually or after significant equipment changes. As a planner, you play a central role in coordinating this effort and making sure that maintenance plans are supported by materials readiness.

25. How do I balance long-term reliability goals with short-term production pressure?

This is one of the most difficult challenges planners face. The key is to build strong relationships with operations and create a shared understanding of the trade-offs involved. Use data to show how deferred PMs or skipped tasks can lead to more frequent breakdowns or higher long-term costs. Where conflict arises, present alternative scenarios - for example, offering multiple downtime windows or splitting tasks into smaller chunks. Develop a schedule that includes contingency space and proactive communication to avoid last-minute changes. Make reliability metrics visible, such as showing how planned work completion reduces emergency maintenance. Over time, consistent, data-backed planning builds credibility. Planners must act as the voice of long-term asset health - not by opposing production needs, but by aligning maintenance priorities with broader business outcomes.

26. What's the role of a planner in a shutdown or turnaround event?

During shutdowns or major turnarounds, the maintenance planner's role becomes even more critical. You're responsible for consolidating scopes from multiple departments, developing detailed job plans, aligning resource requirements, and building the master schedule. This includes managing permits, coordinating contractors, sequencing jobs logically, and confirming materials availability ahead of time. The planner also helps ensure that job durations and dependencies are realistic, and that all execution documents (plans, isolations, RAMS, etc.) are finalised well in advance. On the day, planners often act as coordinators or observers - monitoring progress and adjusting the plan as needed. After the event, capturing lessons learned is essential. In short, your planning rigour is the difference between a smooth, efficient shutdown and one plagued by delays and chaos.

27. How can I keep my backlog under control without constantly deferring work?

Backlog is natural in a healthy maintenance system - but it must be controlled and prioritised. First, ensure all work requests are screened and classified correctly (safety, production impact, non-critical, etc.). Use a backlog review meeting with key stakeholders weekly to challenge each job: is it still required? Can it be grouped with other work? Is there a more efficient way to execute it? Focus on quality over quantity - remove duplicates or vague requests, and convert valid ones into well-scoped work orders. Also, maintain a 4–6 week rolling schedule window so backlog jobs aren't forgotten. Prioritisation frameworks like risk scoring or ABC classifications help manage workload. Remember: backlog isn't a to-do list; it's a dynamic queue that should be constantly cleaned, clarified, and scheduled.

28. How detailed should my job plans really be?

This depends on task complexity, technician experience, and asset criticality. A rule of thumb: the more critical the job, the more detail is justified. For repeatable, simple tasks with experienced staff, a high-level checklist may be enough. But for infrequent or high-risk work, job plans should include clear, step-by-step instructions with required tools, PPE, measurements, photos or diagrams, and estimated time. Think of job plans like an instruction manual - they should enable someone unfamiliar with the asset to complete the task safely and efficiently. Also consider that your job plans are often read by auditors, trainees, or new hires. Over time, build a library of tiered plans - some light-touch, others more comprehensive - and always include a mechanism for technician feedback to continuously improve them.

29. What's the difference between a planner and a scheduler?

Though often blended in smaller teams, the planner and scheduler roles have distinct focuses. A planner defines what needs to be done, how it will be done, what materials are needed, and how long it should take - they build the job plan and get the work order ready. A scheduler looks at when that work should be done, balancing labour availability, operational windows, and prioritisation. The planner ensures jobs are "ready," and the scheduler fits those jobs into a realistic calendar. In some organisations, one person does both. In larger ones, the roles are split. Either way, success depends on close collaboration - poorly planned jobs derail schedules, and poor schedules waste great planning effort.

30. How can I build my confidence as a new planner?

Confidence comes from clarity, preparation, and small wins. Start by understanding your systems deeply - CMMS structure, workflows, how work gets requested and closed. Walk the floor often, ask technicians for feedback, and don't be afraid to say, "I don't know - but I'll find out." Focus on delivering value early: clean up a backlog, improve a PM, or produce a useful report. Each success builds your reputation and belief in your abilities. Also, invest in learning - read industry content, join professional groups, and connect with other planners. Remember, nobody expects perfection right away. What matters is consistency, openness to feedback, and a genuine desire to improve the maintenance process. Over time, your confidence will catch up to your capability.

Conclusion: The Path Ahead

So here we are - at the end of the book, but not the end of your journey.

If you've made it this far, you've already demonstrated something important: you care about doing this role well. You care about planning with precision, growing your skills, improving your team's performance, and building a culture where proactive maintenance wins over chaos.

You've also learned that being a great planner isn't just about CMMS screens and work orders-it's about people. It's about communicating with clarity, solving real problems, influencing without authority, and building trust across your entire operation.

And here's the truth: this role matters more than most people realise.

You are the quiet engine behind uptime. The one who helps make tomorrow more reliable than today. The one who bridges data with action, strategy with execution. You don't wear a cape-but if your planning is done well, a lot of breakdowns are quietly *avoided*. And that is influence.

Keep Growing!

This book has hopefully given you not just a foundation, but a challenge: to keep improving, to keep asking better questions, and to keep making maintenance planning something your organisation respects and relies on.

Whether you're early in your career or a seasoned professional, your growth as a planner never really stops. The technologies might change, the systems will evolve, but the mindset-of adding structure, driving reliability, and supporting your team-that remains your superpower.

Final Words

I wrote this book because I've lived the Maintenance Planning role in a range of settings, a blend of industries, and with a great mix of people, and I know how complex, misunderstood, and undervalued it can sometimes feel. But I also know the satisfaction that comes from bringing order to chaos, seeing plans succeed, and helping people do their best work.

One of the biggest lessons I've learned in the planning role is that influence isn't given - it's earned.

When I first stepped into planning, I thought the technical side would be the hardest part. But it wasn't. It was learning how to bring people on the journey with me - technicians who'd seen it all before, managers with competing priorities, or operators under pressure to hit throughput targets.

What helped? Consistency. Showing up prepared. Following through on what I said I'd do. Being honest when I didn't have the answer, and making sure I figured out what it was.

Over time, I became the person people came to when they wanted things done right. And that, for me, is the quiet power of planning: you don't need a title to lead. You just need to bring value.

So, thank you for reading. Thank you for investing in yourself. And thank you for making our industry better, one plan at a time.

Stay Connected

If you found this book valuable and want to keep growing your skills as a planner, check out Planner HQ - a dedicated growing platform for maintenance planners.

Planner HQ exists to support professionals like you with practical templates, training tools, and insights to help drive reliability, improve planning processes, and grow your impact.

Whether you're looking to deepen your skills, access real-world tools, or connect with other planners, it's a growing hub designed to help you succeed.

Visit www.theplannerhq.com and join the growing community of people just like yourself.

Let's keep learning, improving, and raising the bar for maintenance planning together.

- Chris Diprose

Acronym Buster:

Essential Terms for Maintenance Planners

The world of maintenance and asset management is filled with acronyms - from CMMS to RCA, PdM to ERP. While these terms become second nature over time, they can be a barrier for those new to the field or transitioning into the planning role. This quick-reference guide is here to help. Whether you're brushing up on a term you've forgotten or encountering one for the first time, the following list offers plain-English definitions to keep you confident and informed as you read.

BOM – Bill of Materials

A structured list of all parts, materials, and components required to maintain or assemble an asset or complete a job plan. Helps ensure nothing is missed during planning and execution.

CBM – Condition-Based Monitoring

A maintenance strategy that uses real-time data from sensors and inspections to determine asset health and trigger maintenance only when needed - reducing unnecessary work.

CMMS – Computerised Maintenance Management System

Software used to plan, schedule, track, and record all maintenance activities. It forms the digital backbone of most modern maintenance organisations.

CMRP – Certified Maintenance & Reliability Professional

An internationally recognised certification provided by SMRP to validate knowledge and skills in maintenance and reliability best practices.

EAM – Enterprise Asset Management

A broader system than CMMS, encompassing the full asset lifecycle, from acquisition to disposal, including maintenance, performance, and cost tracking.

ERP – Enterprise Resource Planning

Business-wide software that integrates finance, HR, procurement, supply chain, and sometimes maintenance - often linked to CMMS for seamless data flow.

FMEA – Failure Modes and Effects Analysis

A proactive approach to identify potential failure points in a system or component, assess the impact, and decide how to mitigate those risks before they occur.

HMI – Human Machine Interface

The screen or control panel that allows operators and technicians to interact with machinery or systems, usually displaying performance or fault information.

IAM – Institute of Asset Management

A global professional body that develops and promotes good practices in asset management, including standards like ISO 55000.

ISO 55000 – International Standard for Asset Management

A framework that outlines how organisations can manage the lifecycle of assets systematically and align them with strategic goals.

KPI – Key Performance Indicator

A measurable value used to assess how effectively maintenance activities support organisational goals - e.g., PM compliance, backlog, downtime.

LOTO – Lockout/Tagout

A vital safety procedure that ensures energy sources are isolated and cannot be accidentally restarted during maintenance work.

MTBF – Mean Time Between Failures

A reliability metric that shows the average operating time between equipment failures, helping evaluate asset performance and risk.

MTTR – Mean Time to Repair

An efficiency metric showing the average time required to repair failed equipment and restore it to normal operation.

OEE – Overall Equipment Effectiveness

A composite performance metric combining availability, performance efficiency, and product quality to evaluate how well a machine is utilised.

OEM – Original Equipment Manufacturer

The original producer of a machine or component. OEM manuals and recommendations are often used in job plans and PM development.

PdM – Predictive Maintenance

A maintenance approach that uses data and monitoring technologies to predict failures before they happen, allowing intervention only when needed.

PM – Preventive Maintenance

Scheduled maintenance based on time or usage intervals, intended to prevent failures and extend equipment life before problems occur.

QA/QC – Quality Assurance / Quality Control

Systems and checks that ensure maintenance work is carried out to required standards, specifications, and safety regulations.

RCA – Root Cause Analysis

A structured investigation process that identifies the underlying reason for equipment failure, aiming to prevent recurrence by addressing the real cause.

RCM – Reliability-Centred Maintenance

A methodology used to develop a tailored maintenance strategy by analysing the functions and failure modes of assets.

SCADA – Supervisory Control and Data Acquisition

A centralised system for monitoring and controlling industrial processes, often feeding real-time data into maintenance or automation systems.

SMRP – Society for Maintenance & Reliability Professionals

A global body offering standards, education, and certifications for professionals in the maintenance and reliability industry.

SOP – Standard Operating Procedure

Step-by-step written instructions that describe how to perform a task consistently and safely, often included in job plans or PMs.

Thank you for reading *The Maintenance Planner's Playbook*

We hope this guide helped sharpen your skills, challenge your thinking, and give you practical tools to raise the bar in your planning role.

If you want more templates, tools, and training resources - or want to stay in the loop as we release new guides - head over to:

www.theplannerhq.com

Join the Planner HQ mailing list to:

- Access exclusive downloads
- Get notified about new content, guidebooks, and training tools
- Be part of a growing global community of professional maintenance planners

Your work matters - and better planning makes a real difference.

www.ingramcontent.com/pod-product-compliance
Lightning Source LLC
LaVergne TN
LVHW052113210926
848482LV00025B/576